ION EXCHANGE TRAINING MANUAL

GEORGE P. SIMON

VNR SPRINGER SCIENCE+BUSINESS MEDIA, LLC

Library of Congress Catalog Card Number 90-24735
ISBN 978-94-015-7442-6

16 15 14 13 12 11 10 9 8 7 6 5 4 3 2 1

Library of Congress Cataloging-in-Publication Data

Simon, George P.
    Ion exchange training manual / George P. Simon.
        p.   cm.
    Includes bibliographical references and index.
    ISBN 978-94-015-7442-6   ISBN 978-94-015-7440-2 (eBook)
    DOI 10.1007/978-94-015-7440-2
    1. Ion exchange.   I. Title.
    QD562.I63S55   1991
    541.3'723—dc20   90-24735
                                            CIP

This manual is dedicated to Dr. C. Calmon—
my mentor first, my colleague later, and
unforgettable friend forever.

CONTENTS

PREFACE

It is rare indeed that one comes in contact with a process or technique which impacts many technical disciplines. Ion exchange is such a processs. Although many books have been written on the topic of ion exchange, most have been aimed at the specialist and the graduate engineer or chemist.

The author's experience in ion exchange technology has indicated that there are many specialists in the industry who do not understand ion exchange as a process. Therefore this manual has been written to aquaint and to train. The author has provided background information and hands-on experimental units that can be used to train laboratory technicians who later become assets in the industry. This material has been used by the author for in-house training and at the community college level with success.

It is my sincere hope that the training obtained in this manual will, in some way, be used to improve the environment in which we live. Ion exchange technology has the potential to reduce pollution and improve water supplies when applied properly.

In writing this manual I have had the benefit of valuable assistance. I am indebted to Wes MacGowan and Dr. F. X. McGarvey for helpful suggestions and continued encouragement to get the job done. I have also learned much over the years from Dr. S. Fisher, D. R. Kunin, and Dr. I. Abrams. In one way or another they too have some influence, however indirect, on this modest effort.

GEORGE P. SIMON
GEO-SCI Consulting
Richland, Washington

Part I

HISTORY AND PROPERTIES OF ION EXCHANGE MATERIALS

Chapter 1

ION EXCHANGE
PERSPECTIVES

Ion exchange, as a process, is a natural phenomena, which occurs in soil, minerals, and tissues of both plants and animals. In each instance there are active sites, or groups of sites, which take part in ion exchange reactions. Man has historically utilized these natural phenomena without understanding the mechanism involved. It was in 1850 that Thompson and Way, agricultural chemists, recognized and described the process which has become known as ion exchange (1, 2).

In 1858, Eichorn (3) demonstrated that the ion exchange process defined by Thompson and Way was in reality a reversible process. The first practical industrial application of the ion exchange process occurred in 1905 when Gans (4) synthesized sodium–aluminosilicate cation exchanger materials and used them to soften water. These synthetic inorganic ion exchangers, however, could be used only over a narrow pH range. As a result, only slightly basic natural water supplies could be treated; therefore, the process became known as ''base exchange.''

Early in the 1930s, Smit (5), in The Netherlands, and Liebknecht (6), in Germany, developed cation exchange material from sulfonated coals. These products were found to be stable in dilute acids and were, therefore, applicable over a much wider pH range. Since the sulfonated coals enjoyed a higher stability than the Aluminosilicates in use up to that time, the ion-exchange process found a broader application in the field of water treatment.

In 1935, Adams and Holms (7) in England, developed and patented condensate polymers as structural substrates for various functional groups, thereby creating both anion and cation exchange materials. Cation exchangers were prepared by attaching sulfonic acid groups ($-SO_3^-H^+$) to a phenol formaldehyde polymer matrix. Anion exchangers were prepared by attaching amine groups ($-NH_2$) to a similar matrix. Actually, the anion product was not the strong base anion exchanger available today. Instead, it acted as an acid adsorbent capable of removing only free acids from the process stream. When these two products (cation exchanger in the hydrogen form and anion exchanger in the free base form) were used in series, deionization of water was accomplished for the first time. As a result, the ion exchange process became more versatile. The anion exchange materials (acid adsorbents), developed by Adams and Holms, did not remove weakly dissociated substances, such as carbonic or silicic acids, from processed water. The synthesis of strong base anion exchange materials in 1940 made it possible to remove the weakly dissociated species, and complete demineralization of almost any water supply became a possibility.

These first successful organic ion exchange materials were prepared with condensate polymer structures. These early products had some disadvantages: they were not stable in strong alkalies, and color throw was objectionable, especially at high temperatures.

In 1945, synthetic organic polymers were developed by D'Alelio (8) using styrene and divinylbenzene as the monomers. Divinylbenzene (cross-linker) is a minor constituent (12% or less) but is necessary for the three-dimensional matrix. Both strong base anion and strong acid cation exchange materials could be prepared on the stable polymer matrix. The proper ties of the ion exchange products, in turn, could be influenced by the amount of cross-linking (divinylbenzene) used.

Bauman and coworkers (9) succeeded in preparing these polymers in bead form (up to this time, all ion exchange products were only available in the granular form); while McBurney (11), Wheaton, and others, demonstrated that the bead size could be controlled within a narrow range. In addition, it was McBurney (11) and the Wheaton–Bauman team (10) who successfully prepared strong base anion exchange materials on these bead-form polymeric structures.

Commercial cation and anion exchange products prepared on the copolymers have been successfully applied to water treatment and other industrial processes since the 1940s. These products have become known as "gel" type ion exchange materials. While these gel-type ion exchangers were very successful (12), they were observed to become fouled and almost inoperative in certain applications where surface water supplies were being treated. In 1950, the foulant was identified as high molecular weight organic acids (Humic and Fluvic acids), commonly found in surface water supplies. These acids (molecular weights of 100,000 or more) were found to be irreversibly exchanged on strong base anion exchange materials. Pretreatment systems were developed to eliminate the fouling in these 10 applications; however, the attempt met with mixed success.

In an attempt to overcome this irreversibility in the ion exchanger, Corte (14) and Meitzner and Oline (15) prepared anion exchange materials with large pores to improve the internal diffusion rates for large molecules. These became known as macroporous, macroreticular, fixed pore or iosporous ion exchangers depending on the manufacturer.

Also during the 1950s, Juda and McRae (13) developed ion exchange products in the form of membranes. The cation exchange membrane product allows the passage of cation under the influence of an electric current, and the anion membrane product allows the passage of anions. By arranging these materials in a series of cells (known as a *stack*), it is possible to deionize water or treat industrial wastes by using an electric current without employing chemical regenerants.

In spite of the relatively high porosity, compared to the "gel" exchangers, strong base anion exchanger materials were still sensitive to organic fouling. Part of this was due to the fact that industry was demanding better and better performance from these products. Initially, these fouling problems were reduced by improvements to the pretreatment systems and periodic resin-cleaning procedures. It was determined that the basic cause for this type of fouling was that the organic matrix had a strong adsorption affinity for the humic and fluvic acids in certain surface water supplies. The adsorption appeared to be almost irreversible in character. In 1952, ion exchange materials based upon acrylic polymers were developed and patented by H. Schneider (16). When these products were applied to the treatment of surface waters containing organic foulants, it was observed that the organics were more efficiently removed during the regeneration process.

This is a relatively short review of the development of ion exchange, leading up to modern ion exchange materials available commercially, and covers a time span of about 140 years.

The relationship between exchange product development and ion exchange process development is clearly evident.

At the present time, there are four manufacturers of ion exchange products in the United States, and an equal number of major overseas manufacturers.

As expected, each manufacturer has its own trademark, name, and, of course, its own product-by-product designations. Thus, it is often confusing when one tries to find equivalent ion exchangers from trade names or product numbers alone.

In addition, trade literature often leads one to believe that one manufacturer's product is somehow better than any other. From bead quality and other physical properties, an ion exchanger's chemical properties and operating performance is dependent, for the most part, on the functional group; and to a lesser extent on the matrix that supports that group.

Therefore, it is possible to classify the diversity of products into equivalent groups based on the active group or groups. This type of comparison is provided in Table 1-1. Detailed information for these ion exchange products can be obtained by writing or calling the manufacturers directly. Addresses and the phone numbers (when available) have been provided as footnotes to Table 1-1.

Applications in *Ion Exchange* Technology

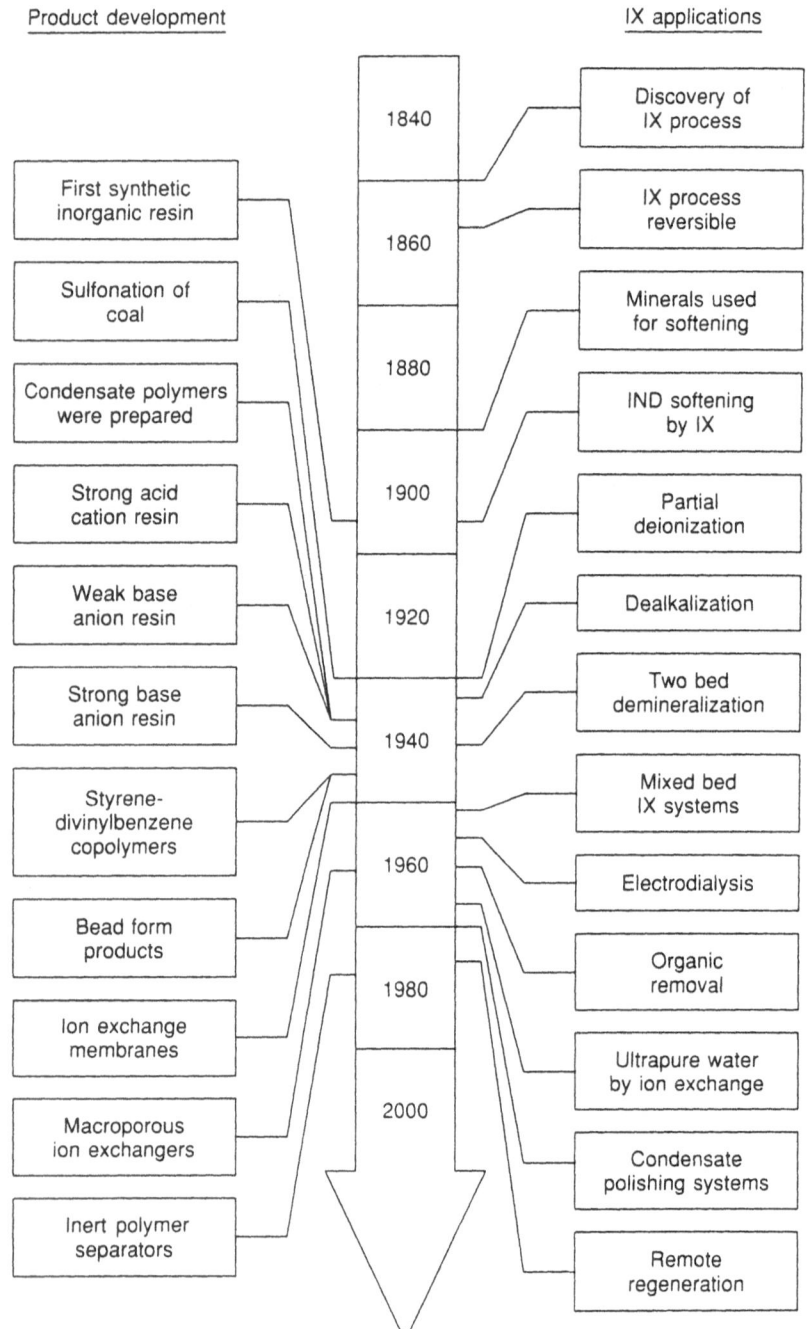

Figure 1-1. Historical development of products and applications in ion exchange technology.

Table 1-1. Comparison of Equivalent Commercial Ion Exchange Products

Product classification			U.S. manufacturers					Overseas manufacturers		
Matrix	Generic type	Active group(s)	Amberlite[a]	Dowex[b]	Ionac[c]	Purolite[d]	Diaion[e]	Imac[f]	Lewatit[g]	Wofatit[h]
Styrene-DVB gel-type polymers	Strong acid cation	$R-SO_3^-$	IR-120 IR-122 IR-123	HCR HCR-S HGR HCR-W HDR	C-240 C-298 C-250 C-299	C-100	SK 1B SK 100	C 12 C 8P	S 100 S 115	KPS 200 F P
	Weak acid cation	$R-COOH$	IRC-50 IRC-77 IRC-84 C-433 C-464	CCR-2	C-270 CC	C-105		Z 5	CNP	CP 300 CN
	Strong base anion Type I	$R-N(CH_3)_3^+$	IRA 400 IRA-402 A-104	SBR SBR-P 11	ASB-1 ASB-1P A-540	A-600 A-400	SA 10A SA 12A	S 5-40 S 5-50	M 600	N L 165
	Strong base anion Type II	$R-N\!\!\begin{array}{l}C_2H_5OH\\(CH_3)_2^+\end{array}$	IRA 410	SAR	ASB-2 A-550	A-300	SA 20A		M 500 M 504	MD L 150
	Weak base anion	$R-N(R)_2$ $R-NHR$ $R-H$ $R-H(R)_2$	IRA-45 IRA-47 IRA-68	WGR WGR-2	A-300 A-315	A-845		A 20 A 21 A 27		
Styrene DVB macroporous polymers	Strong acid cation	$R-SO_3$	200 252	MSC-1	CFS CFP-110 C-360	C-150	PK 216 PK 228	C 16P	SP 120	
	Weak acid cation	$R-COOH$	IRC-77 IRC-718 DP-1		CNN CC	C-106 NRW-100	WK 10 WK 30			
	Strong base anion Type I	$R-N(CH_3)_3^+$	IRA-900 IRA-904 IRA-938	MSA-1	A-642 A-641 A-741	A-500 A-500P A-501P	PA 308 PA 312 PA 316		MP 600	

Table 1-1. (Continued)

Product classification			U.S. manufacturers					Overseas manufacturers		
Matrix	Generic type	Active group(s)	Amberlite[a]	Dowex[b]	Ionac[c]	Purolite[d]	Diaion[e]	Imac[f]	Lewatit[g]	Wofatit[h]
	Strong base anion Type II	$R-N\genfrac{}{}{0pt}{}{C_2H_5OH}{(CH_3)_3{}^+}$	IRA-910 IRA-911	MSA-2	A-651	A-510	PA 416	S 5-42	MP 500	
	Weak base anion	$R-N(R)_2$ $R-NHR$ $R-NH$ $R-H(R)_2$	IRA-93 IRA-94	MWA-1	AFP-329	A-100	WA 20 WA 30		M 62 M 64	
Acrylic polymers	Strong base anion Type I	$R-N(CH_3)_3{}^+$	IRA-458							
Mixed Beds	Strong acid and strong base		MB-1 IRN-150		MR-3	NM-60 NM-65 NM-75	NRW-37			
	Strong acid and weak base		MB-4							

[a] ROHM and HAAS Co., Independence Mall West, Philadelphia, PA 19105, (215) 592-3000 or (800) 523-0762.
[b] Dow Chemical Co., 2040 Willard H. Dow Center, Midland, MI 48674, (517) 636-2286.
[c] SYBRON Chemicals, Inc., Birmingham Rd., PO Box 66, Birmingham, NJ 08011, (609) 893-1100 or (800) 678-0200.
[d] The Purolite Co., 150 Monument Rd., Bala Cynwyd, PA 19004, (215) 668-9090 or (800) 343-1500.
[e] Mitsubishi Chemical Ind., Ltd., No. 4 2-Chome, Marunouchi Chiyoda-Ky, Tokyo, Japan; also Dianex Systems, Inc., 231 South Transit St., Lockport, NY 14094, (818) 914-1000.
[f] Imacti, Postfach 240, Amsterdam-C, The Netherlands.
[g] Farbenfabriken Bayer, Leverkusen, Germany.
[h] VEB Flarbenfabrik Wolfen, Wolfen Kr., Bitterfield, Germany.

REFERENCES

1. H. Thompson, Adsorbent power of soils. *J. Roy. Soc.* 11, 68 (1850).
2. J. T. Way, Power of soils to absorb manure. *J. Roy. Soc.* 11, 313 (1850); *J. Roy. Soc.* 13, 123 (1852).
3. H. Eichorn, On the reactions of silicates with dilute solutions of salts. *Pogendorf's Ann. Phys. Chem.* 105, 126, (1858).
4. R. Gans, Zeolites and similar compounds: Their construction and significance for technology and agriculture. *Jahrb. preuss. geol. Landesanstalt* 26, 179 (1905); 27, 63 (1906); *Zentr. Mineral Geol.* 22, 699, (1913); German patents: 174,097 and 197,111 (1906); U.S. patents: 914,405, (1905), 943,535 (1906), 1,131,535 (1915).
5. P. Smit, U.S. patents: 2,191,063 (1940); 2,205,635 (1940).
6. O. Liebkneckt, U.S. patents: 2,191,060 (1940), 2,206,007 (1940).
7. B. A. Adams and E. L. Holms, Adsorptive power of zeolites. *J. Soc. Chem.* 54, (1935).
8. G. T. D'Alelio, U.S. patents: 2,340,110 (1944), 2,340,111 (1944), 2,366,007 (1945), 2,366,008 (1945).
9. W. C. Bauman and J. Eichorn, *J. Am. Chem. Soc.* 69, 2380, (1947).
10. R. M. Wheaton and W. C. Bauman, *Ind. Eng. Chem.* 45, 228 (1947).
11. C. H. McBurney, U.S. patent: 2,591,573 (1952).
12. R. Kunin and F. X. McGarvey, *Ind. Eng. Chem.* 43, 734 (1951).
13. W. Juda and W. A. McRae, *J. Am. Chem. Soc.* 72, 1044 (1950).
14. H. Corte, et al., German patent: 1,113,576 (1957).
15. E. F. Meitzner and J. A. Oline, French patent: 1,237,342 (1960).
16. H. Schneider, U.S. patent: 2,679,359 (1952).

Chapter 2

SYNTHESIS OF ION EXCHANGE MATERIALS

INTRODUCTION

The development of ion exchange materials, from the natural minerals to the synthetic products in wide use today, was driven by two basic needs:

- To provide higher operating capacities and better product quality-control.
- To provide those products which anticipate, or solve municipal and industrial water-treatment problems.

In the field of water-treatment technology, process modifications often affected product development, and new products often opened new approaches in process technology.

NATURAL ION EXCHANGE MATERIALS

Considering the historical review in Chapter 1, it is not surprising that the first ion exchange material applied to commercial processes was derived from natural minerals. The minerals exploited were found in deposits known as marls or greensands. Gluaconite, for example, has a very complex composition [(K, Na) (Al, Fe, Mg)$_2$ (Al, Si)$_4$O$_{10}$ (OH)$_4$].

The chemical compositions, which are also complex, of a few of these natural ion exchange minerals are given in Table 2-1.

All natural materials of this kind appear to form in lens-shaped cavities associated with ancient lava flows, probably a late deposit from water which invaded the lava strata. While these natural minerals were useful (some deposits were actively worked in the United States as late as 1950), they were difficult to purify for commercial use. Product quality and uniformity were highly variable quantities. In addition, these materials had a relatively narrow pH range in

Table 2-I.   Composition and Capacities of Some Natural Ion Exchange Minerals

General type	Mineral name	Composition (formula)	Capacity (meq/dry gm)
Fibrous	Edingtonite	$Ba(Al_2Si_3O_{18})4H_2O$	3.9
	Natrolite	$Na_2(Al_2Si_3O_{18})2H_2O$	5.3
	Scolecite	$Ca(Al_2Si_3O_{16})3H_2O$	5.0
Lamellar	Stilbite	$(Na,Ca_5)(Al_2Si_3O_8)3H_2O$	3.2
	Heulandite[a]	$Ca(Al_2Si_6O_{16})5H_2O$	3.3
	Montmorillonite	$(Ca,Mg)O(Al_2Si_3O_{13})6H_2O$	1.5
Cubic	Analcite[a]	$Na(AlSi_2O_6)H_2O$	4.5
	Mordenite	$(Ca_5Na)(AlSi_5O_{12})1/3H_2O$	2.3
	Chabazite[a]	$(Ca_5Na)(AlSi_2O_6)3H_2O$	4.0
	Faujasite	$(Ca,Na_2)(Al_2Si_5O_{10})6.6H_2O$	3.9
	Mordenite	$(Ca_5,Na,K)(AlSi_{10}O_{23})7H_2O$	0.89
Feldspars	Leucite	$K(AlSi_2O_6)$	4.6
	Sodalite[a]	$Na(AlSiO_4)2/3H_2O$	9.2
	Ultramarine	$Na(AlSiO_4)1/3H_2O$	8.3
	Clinoptilolite[a]	$[(Ca,Na_2)(AlSi_7O_{16})]6H_2O$	1.85

[a]These specific minerals are often referred to collectively as natural zeolites.

application. Below pH 6.0, they tend to breakdown or dissolve in use; and above 8.0, the operating performance decreased. Clinoptilolite, a selective zeolite for ammonia removal, is still available commercially.

SYNTHETIC ALUMINOSILICATES

Synthetic aluminosilicate ion exchangers were developed to provide products which were consistent from batch to batch, and lower in production cost. Initially, these synthetic products were prepared by high-temperature (sintering) trechniques, and the glass-like product was ground and washed before use. Later a product known as DECALSO was developed by using solutions of alum, sodium aluminate, and sodium silicate to form a hydrous gel.

This gel was separated from the excess water by filtration in a plate and frame filter press, dried and crushed to a convenient size for commercial applications.

These synthetic aluminosilicate ion exchangers were three times the capacity per cubic foot than the natural minerals then in use. Therefore, many of the existing processes (i.e., softening) were converted to this new product. While better than the natural minerals, and more economical to use, these aluminosilicates had the same pH disadvantages characterized by the natural ion exchange materials.

SULFONATED COALS

Here again, the problems of the past dictated the direction of product development. The sensitivity to high and low pH of the synthetic aluminosilicate ion exchangers excluded these products from many applications of interest.

Early in the 1930's, both in The Netherlands (Smit) and in Germany (Liebbknecht), sulfonated coals were developed and used in commercial applications as ion exchange media. Sulfonated coal was the first cation exchange product that was stable in dilute acids and was easily regenerated to the hydrogen form with dilute acids (i.e., 2% H_2SO_4). One of the first applications of sulfonated coal, aside from softening, was dealkalization (discussed in Chapters 4 and 5). Dealkalization actually reduced the ion content of the water being treated, and therefore was the first clear demonstration that relatively pure water could be produced by ion exchange processes. It appeared that the ion exchange process had to be less costly than conventional distillation. While partial deionization was achieved, it was evident that complete deionization could not be accomplished until a product was developed which was capable of removing all the anions associated with the cations in natural water supplies.

CONDENSATE POLYMER PRODUCTS

In 1935, Adams and Holmes developed the first synthetic organic anion and cation exchange materials. These English chemists used an insoluble polymer prepared from phenol and formaldehyde as the basic structure, and attached appropriate active groups to provide cation exchange and anion exchange materials. These ion exchange products were available as granular exchangers. The cation exchanger was prepared by using a mixture of phenol, phenolsulfonic acid, and formaldehyde as shown in Figure 2-1. A similar process scheme yields a weak base anion exchanger as shown in Figure 2-2. When these regenerated ion exchangers were used to treat water, it was observed that almost complete deionization could be achieved, as indicated by the following reactions:

$$
\begin{array}{cccc}
 & \text{Regenerated} & \text{Exhausted} & \\
 & \text{cation} & \text{cation} & \text{Cation} \\
\text{Influent water} & \text{resin} & \text{resin} & \text{effluent} \\
\begin{bmatrix} NaCl \\ CaSO_4 \\ Mg(HCO_3)_2 \\ SiO_2-H_2O \end{bmatrix} & + R-SO_3H \rightleftharpoons & \begin{bmatrix} R-SO_3Na \\ (R-SO_3)_2Ca \\ (R-SO_3)_2Mg \end{bmatrix} & + \begin{bmatrix} HCl \\ H_2SO_4 \\ CO_3 \\ SiO_2-H_2O \end{bmatrix}
\end{array}
$$

$$CO_2 \text{ Removal by degasification}$$

$$
\begin{array}{cccc}
 & \text{Regenerated} & \text{Exhausted} & \text{Treated} \\
\text{Cation effluent} & \text{anion resin} & \text{anion resin} & \text{water} \\
\begin{bmatrix} HCl \\ H_2SO_4 \\ SiO_2-H_2O \end{bmatrix} & + R-N(CH_3)_2 \rightleftharpoons & \begin{bmatrix} R-N(CH_3)_2-HCl \\ R-N(CH_3)_2-H_2SO_4 \end{bmatrix}_{SiO_2-H_2O}
\end{array}
$$

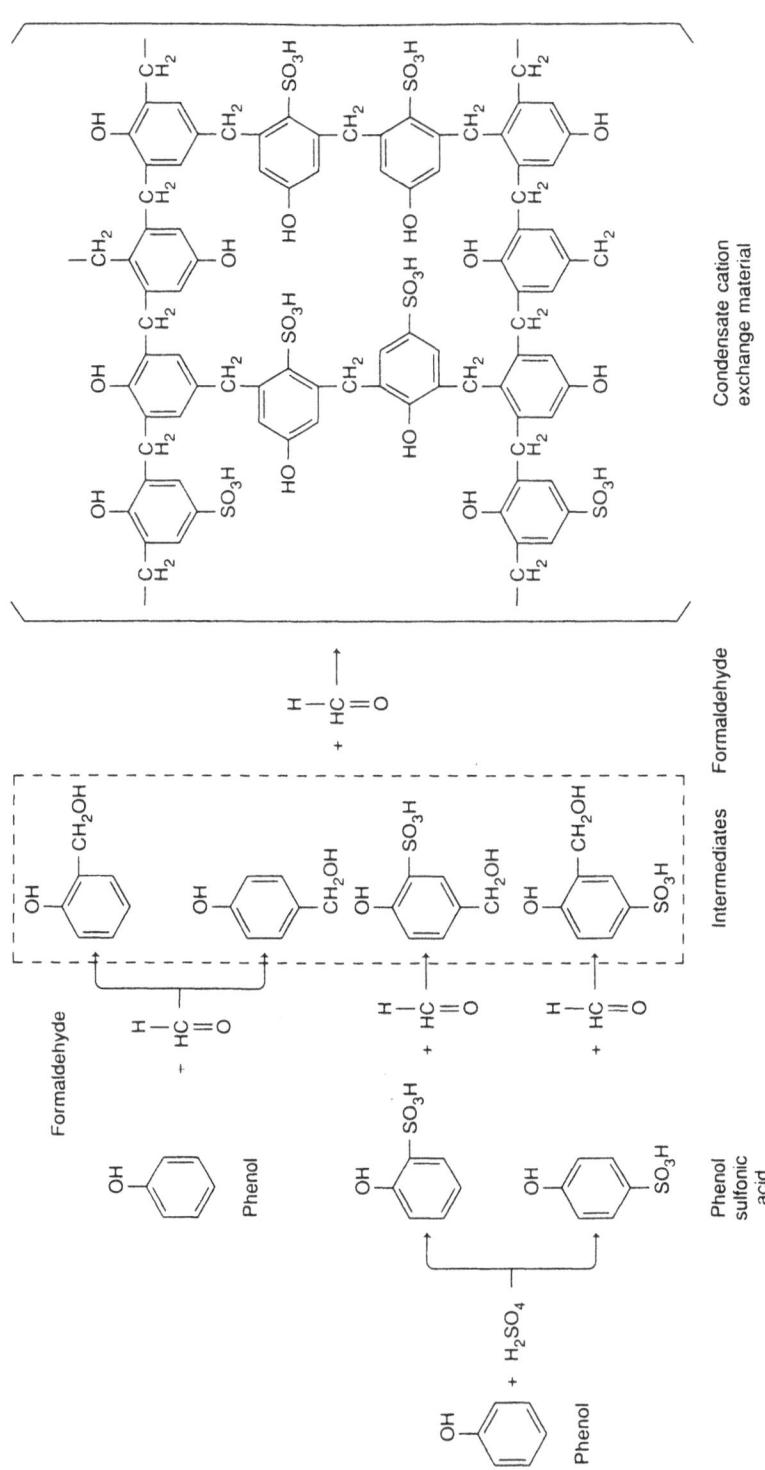

Figure 2-1. Preparation of a condensate strong acid cation exchanger.

13

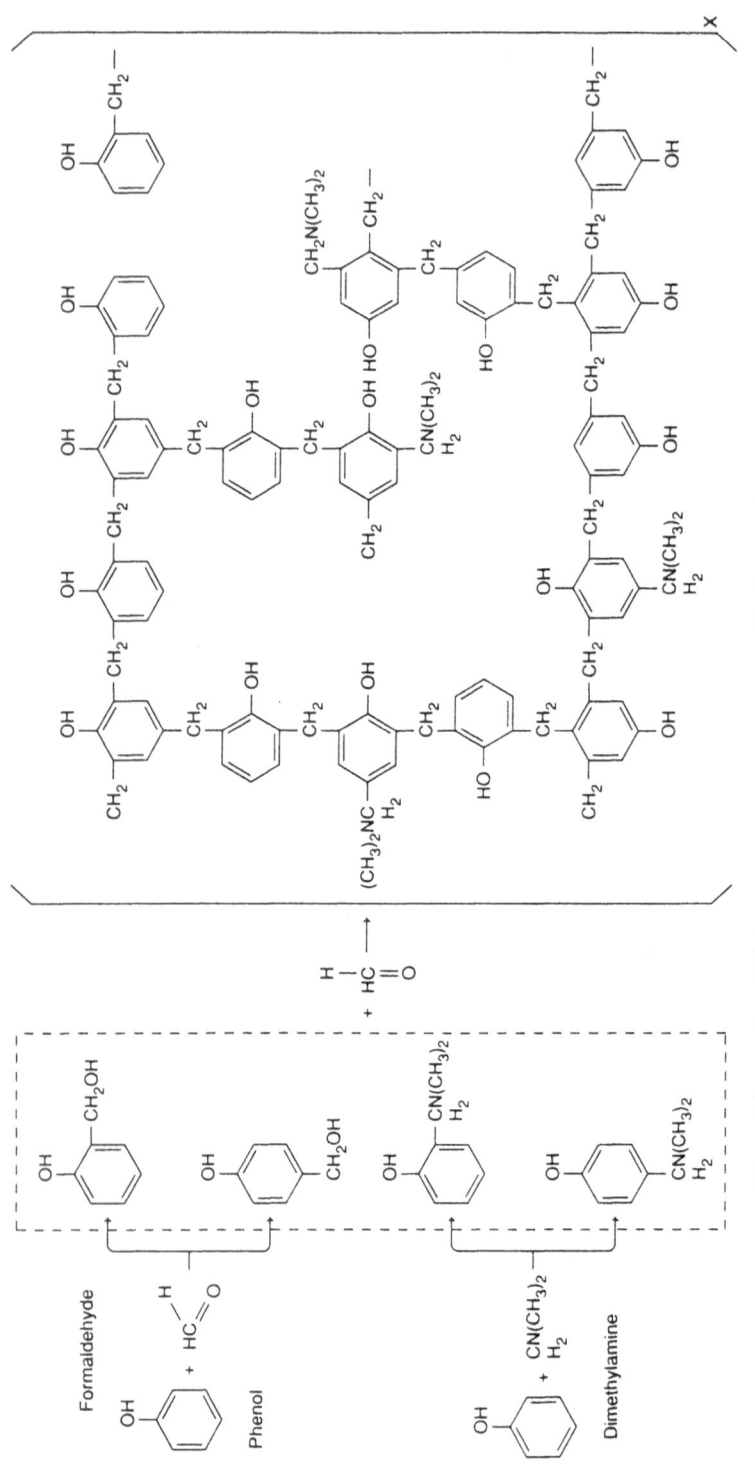

Figure 2-2. Preparation of a condensate weak base anion exchanger.

14

Where, R, represents the insoluble condensate polymer matrix. Residual silica was the only weakly dissociated anion which appeared to prevent the production of pure demineralized water at low cost. Silica removal was attempted by pretreating the water supply with HF to convert the SiO_2 to fluorosilicic acid (H_2SiF_6) and by removing it with the weak base anion exchange material. Although the process worked, its commercial application was limited.

At about this time, electrical utilities were increasing steam-generator pressures and capacities to improve plant operating efficiencies. Conventional boiler feedwater treatment systems were unable to produce the low silica residuals (i.e., <1 ppm) required by these improvements. As a result, not surprisingly, development of anion exchange materials continued in an effort to:

- Provide products which could be used to prepare large volumes of demineralized water to satisfy growing industrial demands.
- Produce uniform products with properties which provide broader application possibilities.

POLYMERIC ION EXCHANGE PRODUCTS

A timely development occurred in 1945 when D'Alelio patented a three-dimensional solid by copolymerization of styrene and divinylbenzene. This polymerization is shown schematically in Figure 2-3.

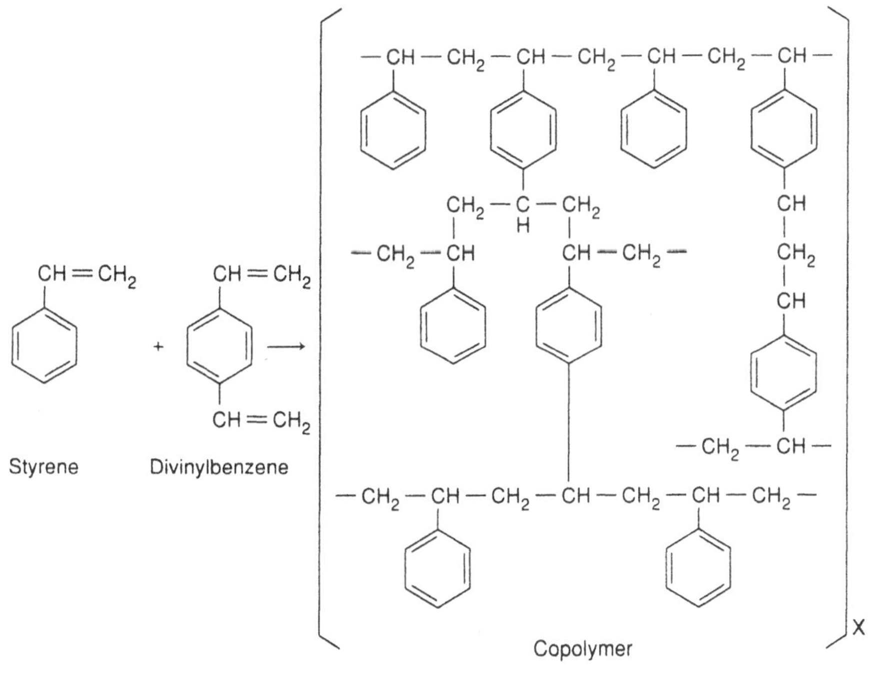

Figure 2-3. Polymerization of styrene and divinylbenzene.

These initial materials were prepared in bulk, and it became apparent that the finished product properties could be controlled by varying the amount of divinylbenzene cross-linking used.

In 1947, Bauman and Eichorn succeeded in their efforts to prepare the copolymer in the form of beads, and demonstrated that the product screen size could be controlled within a narrow range (i.e., 10–40 mesh). The first ion exchange material using this unique copolymer was a strong acid cation exchanger. The preparation of this ion exchanger is illustrated in Figure 2-4.

McBurney, Wheaton, and Bauman continued these studies and successfully prepared strong base anion exchangers on the 4% cross-linked polymer. The preparations of the chloromethylated, intermediate, and the finished strong base anion exchangers are shown in Figures 2-5, 2-6, and 2-7, respectively.

The importance of this development was immediately recognized, and the new technology was applied to preparation of weak acid cation exchangers (Figure 2-8) and weak base anion exchangers (Figure 2-9).

Thus, the products prepared from this polymer became the backbone of the ion exchange industry. With the strong base anion exchangers, silica residuals could be routinely reduced to ppb concentrations. In addition, high purity water, equal to distilled water, could be produced in bulk at a fraction of the cost of conventional distillation. The higher capacities achieved with these new products made them more economical, lowered operating costs, and reduced the vessel size. These materials were more versatile and brought about a wide application in the industrial arena. This same versatility brought additional problems to the surface, as the ion exchangers were used to treat more and more water supplies. For example, irreversible fouling was observed when these gel-type anion exchangers were used to treat water supplies containing low concentrations of natural organic material.

POROUS ION EXCHANGE PRODUCTS

This organic fouling was found to reduce the anion exchanger performance and service life. Initially, it was apparent that pretreatment of the suspect water supplies was required for special applications.

It wasn't until the 1950s that the foulant was identified as a high molecular weight humic acid common to many surface waters, which was found to exchange irreversibly on strong base anion exchangers. To facilitate the diffusion of these large molecules in and out of the ion exchanger beads, newer exchangers were prepared with built-in pores. These, second generation ion exchange products are referred to as macroporous, macroreticular, or fixed-pore ion exchangers, depending upon the manufacturer.

In spite of the relatively high porosities, some strong base anion exchangers with aromatic polymer matrices were still observed to foul in some water-treatment applications.

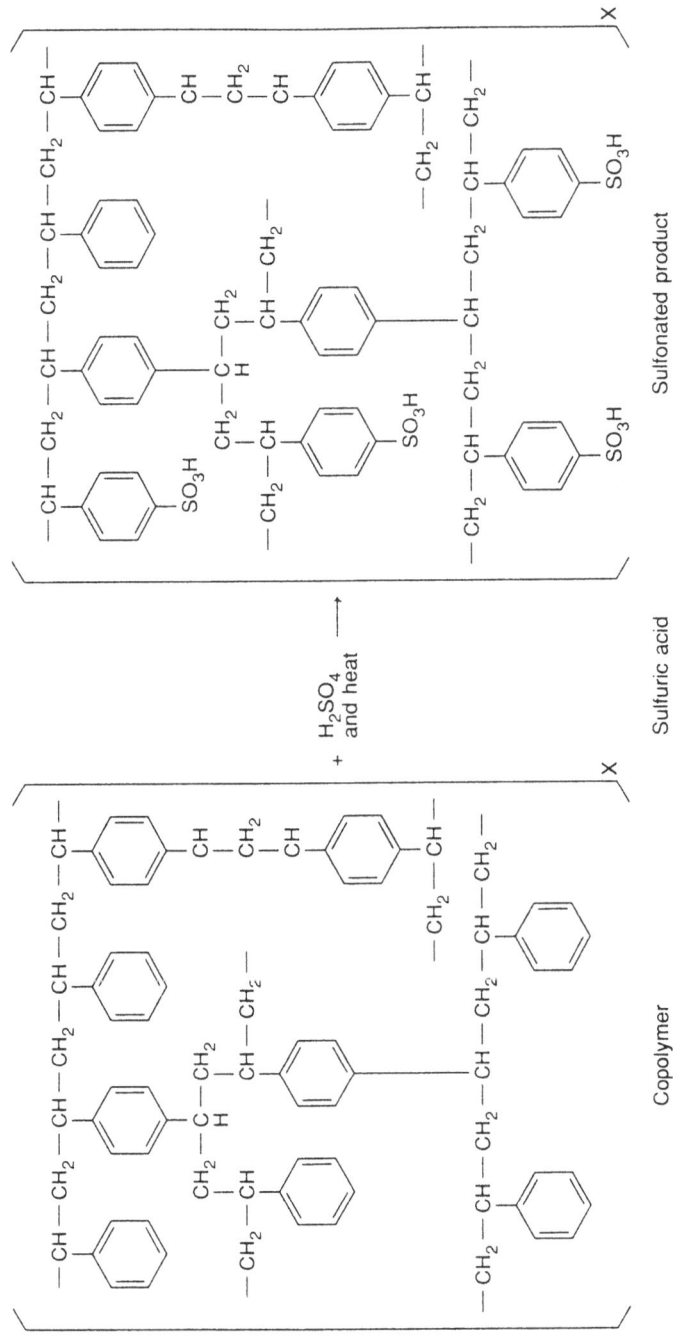

Figure 2-4. Sulfonation of styrene–divinylbenzene copolymer.

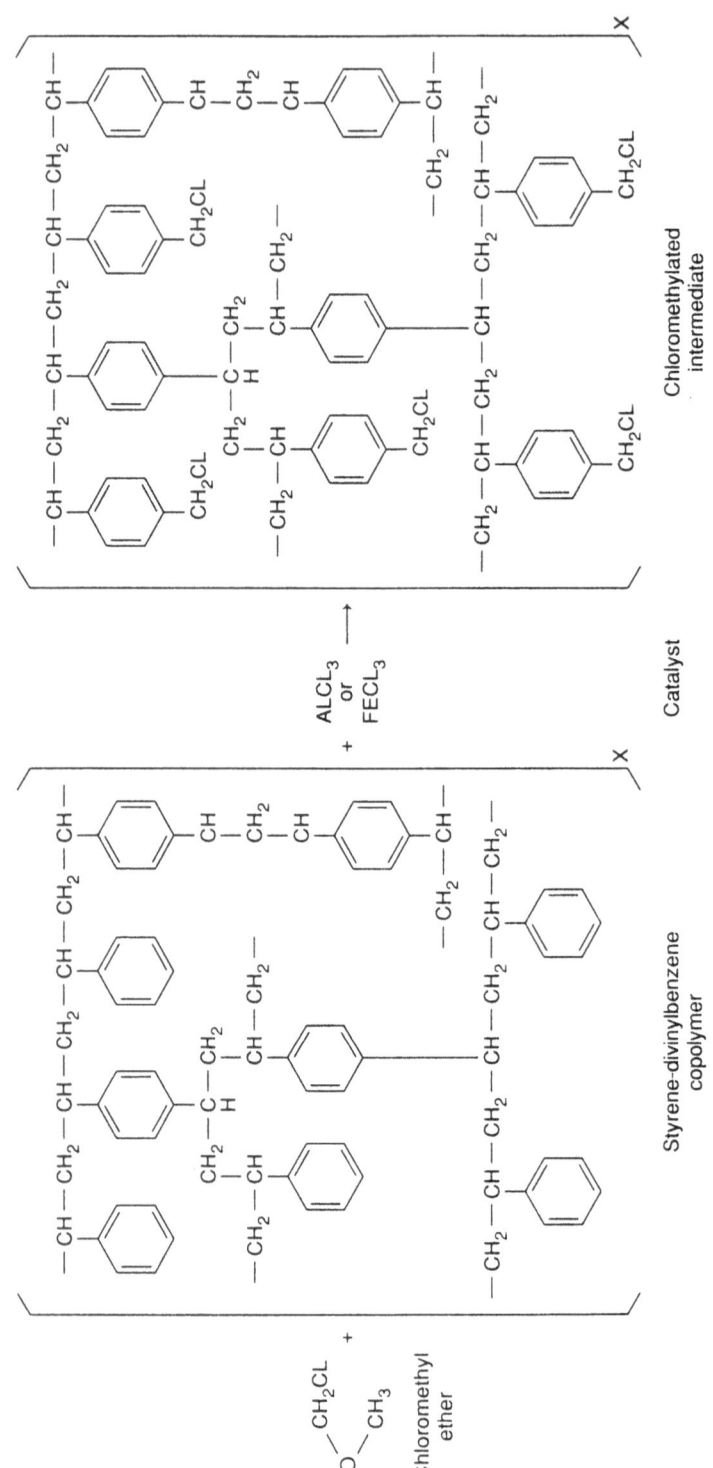

Figure 2-5. Chloromethylation of styrene-divinylbenzene copolymer.

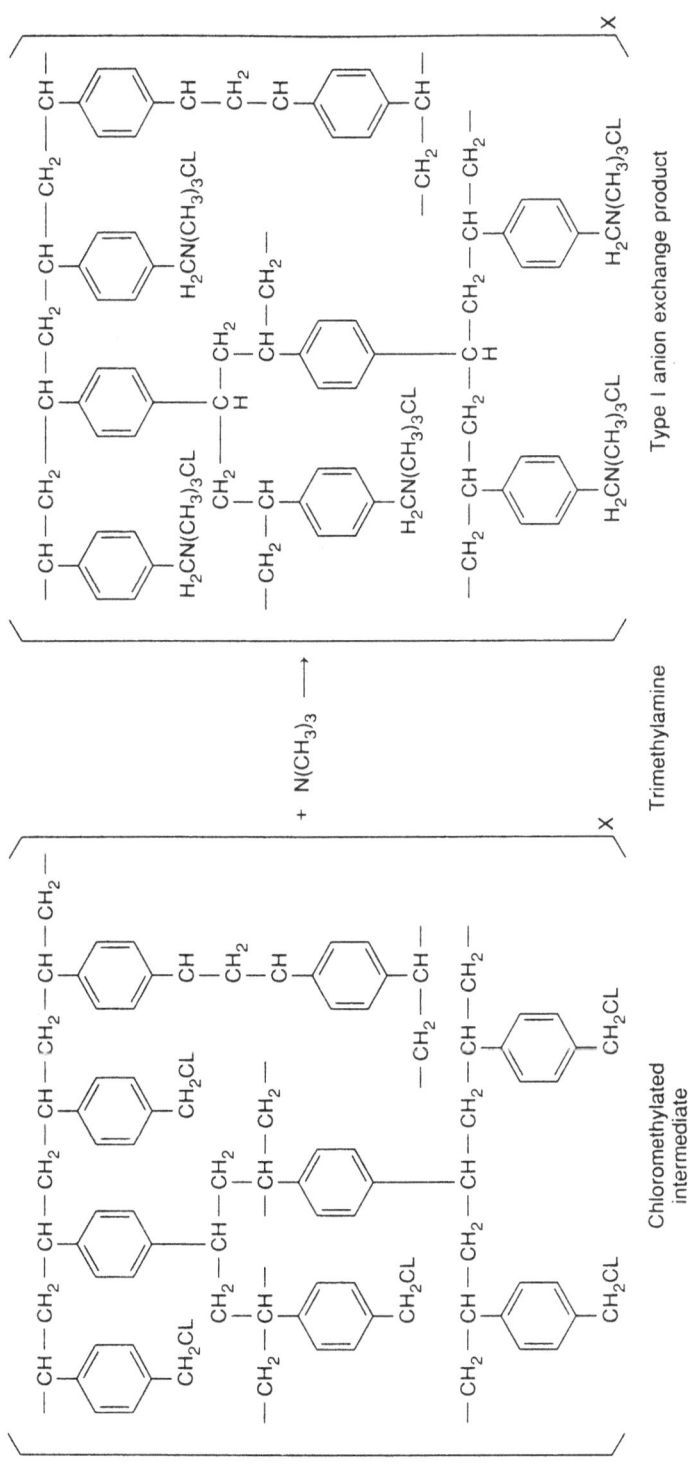

Figure 2-6. Preparation of a type-I strong base anion exchanger from the chloromethylated intermediate.

19

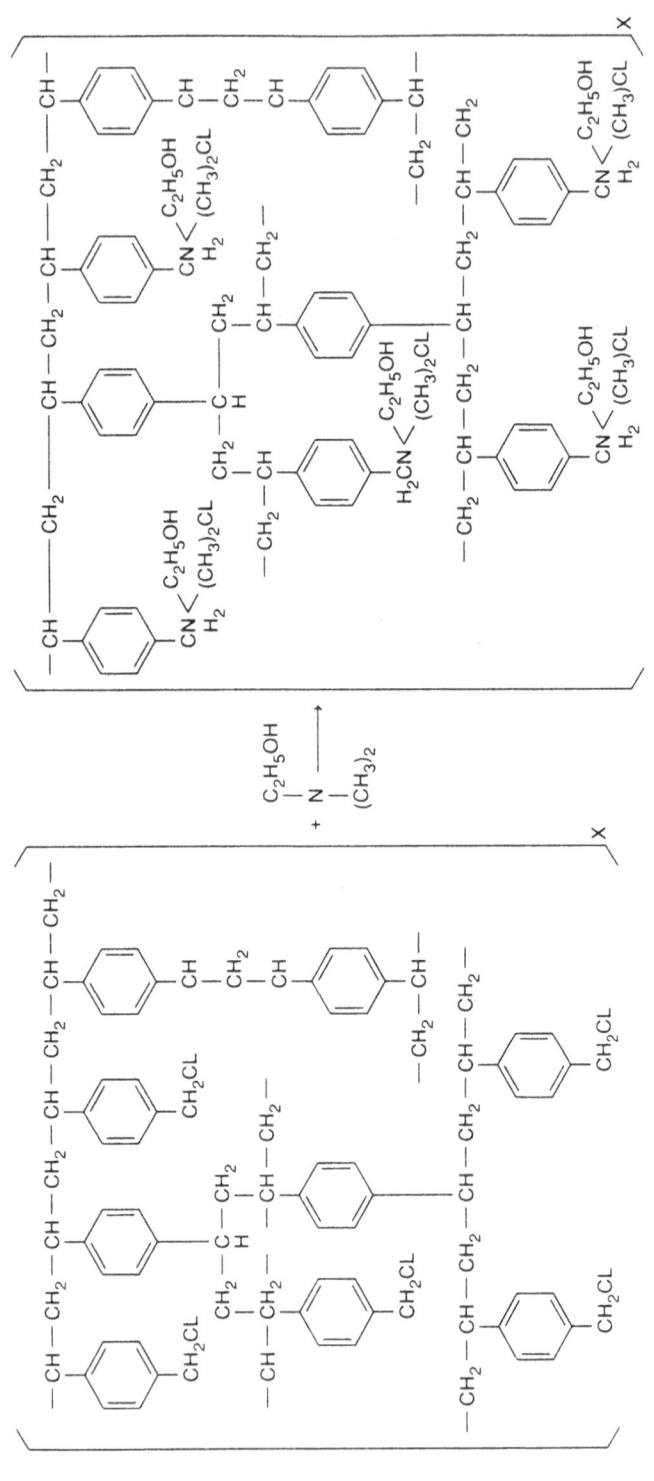

Figure 2-7. Preparation of a type-II strong base anion exchanger from the chloromethylated intermediate.

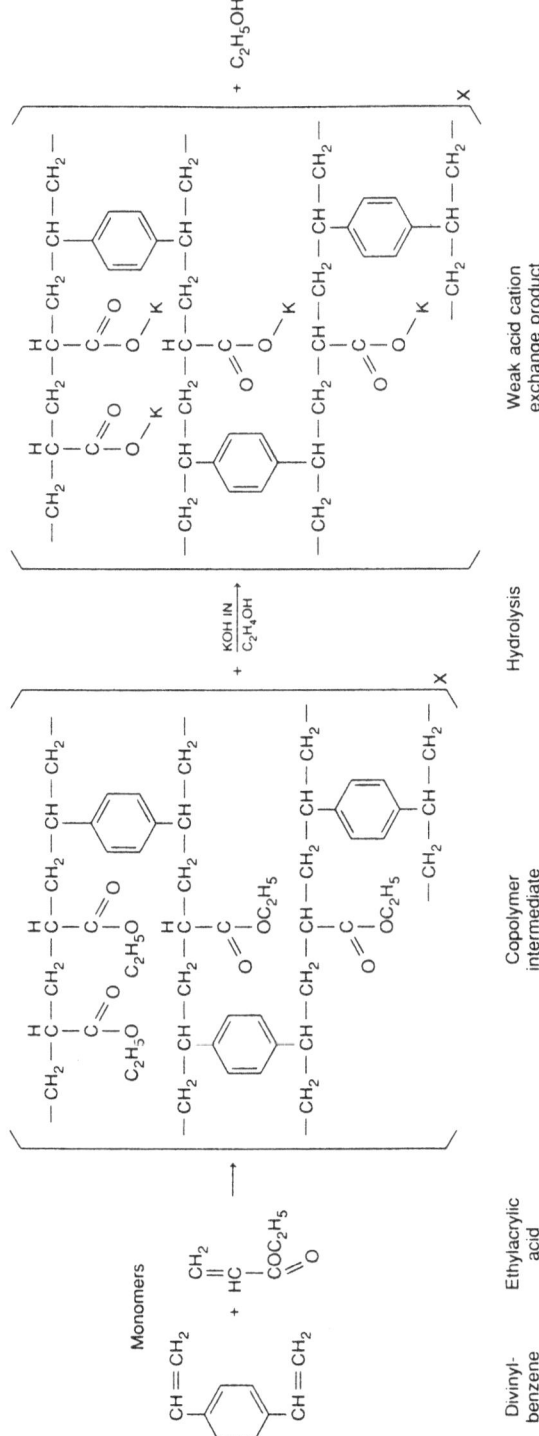

Figure 2-8. Preparation of a weak base anion exchanger from chloromethylated intermediate.

21

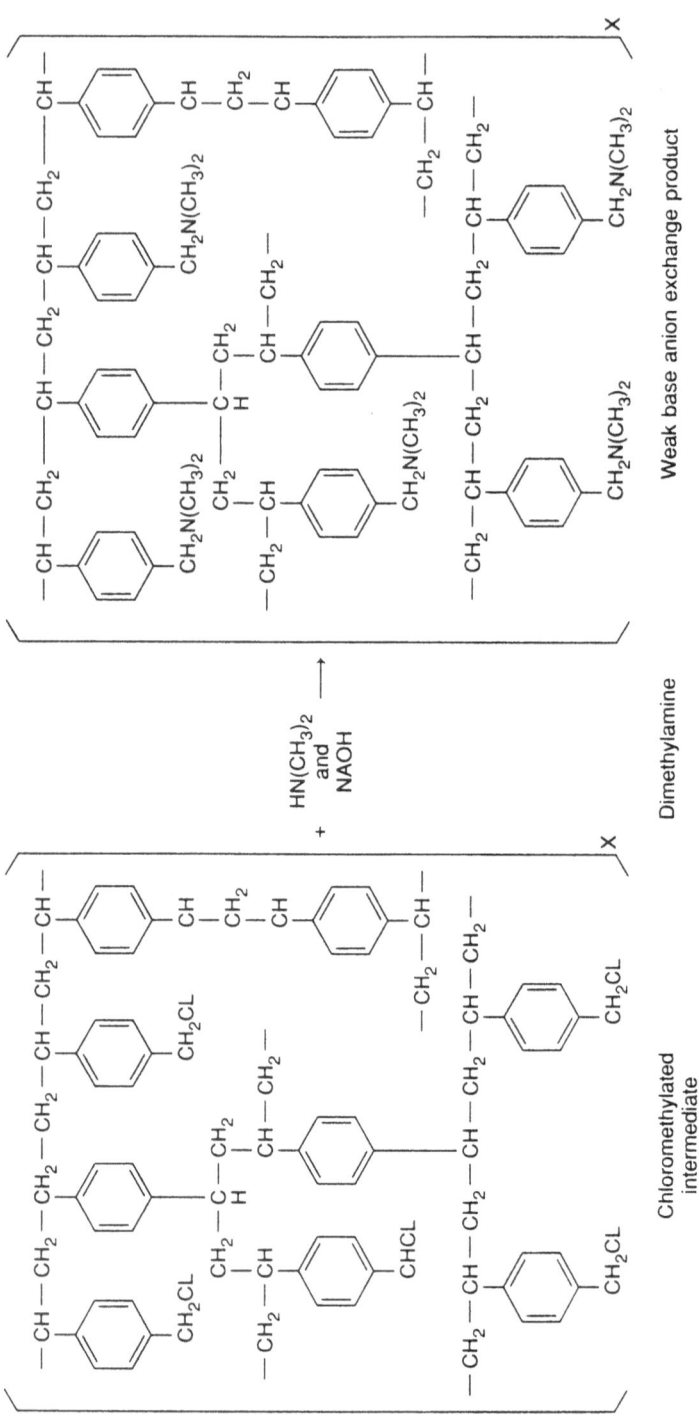

Figure 2-9. Preparation of a weak acid cation exchanger.

Operational problems of this kind were solved, at first, by developing better pretreatment and resin-cleaning processes. The basic cause of the problem was found to be associated with the aromatic polymer structure. These polymers exhibited a strong affinity for the humic acids present in some water supplies. As a result, irreversible absorption occurred and the foulants were not released efficiently during normal regeneration procedures.

Ion exchange materials with aliphatic structures were then developed using acrylic acid as a starting monomer. When these new products were used to treat the identified problem water supplies, it was observed that the adsorbed foulants could be removed more efficiently during normal regeneration processes.

Although the acrylic-based anion exchangers were developed in 1952, their importance and commercial application in water treatment was not realized until the mid 1970s.

ION EXCHANGE MEMBRANE PRODUCTS

These products are ion exchange materials fabricated in sheet form rather than the conventional bead form. While the ion exchange membranes share a number of common properties with the bead-form products, there are some noteworthy differences.

- Ion exchange membranes can be employed continuously instead of in a batch-operating mode.
- They can be used to process both dilute and concentrated solutions, and are more efficient when treating concentrated solutions.
- Since electric current is the driving force, membrane processes do not produce waste chemical regenerants and therefore will have a smaller effect on the environment.

The ion exchange membrane products are available in two basic forms, and each has somewhat different physical properties as described below.

Homogeneous Membrane Products

Homogeneous ion exchange membrane materials may be considered to be three-dimensional networks of an insoluble organic polymer matrix. The active groups such as $-SO_3H$, $-(CH_3)_3$, or $-COOH$ are chemically attached to the three-dimensional matrix. Once the spaces in the matrix are filled with water, the attached reactive groups are free to dissociate in a manner similar to bead-form ion exchangers, and are capable of exchanging any ions which may be in the interstitial water phase.

Heterogeneous Membrane Products

Heterogeneous ion exchange membranes are manufactured by depositing finely ground anion or cation exchange materials in an inert fiber sheet. The ion exchange particles are then fixed by using a binder and applying pressure and heat. This technique yields a membrane product that generally has better dimensional stability; however, the process does result in a product with somewhat lower ion exchange capacity.

Ion exchange membrane products can be considered as three-dimensional networks (sheets) prepared of an insoluble polymer matrix. Attached to the network chains are the reactive groups which give the membrane its chemical characteristics. Cation-permeable products will have $-SO_3H$ groups and anion-permeable products usually have $-N(CH_3)_3OH$ groups. When wet, the spaces between the polymer chains are filled with water, and the reactive groups are completely hydrated and dissociated. Ideally, the random passages are large enough for most ionic species to pass through by displacing the co-ions attached to the reactive groups. When $-SO_3H$ groups are present only cations will be transported by an electric current. On the other hand, the $-N(CH_3)_3OH$ groups allow only anions to pass easily. As a result, the products are said to be either "cation-selective" or "anion-selective." The word *permselectivity* has been coined to describe this basic property of ion exchange membranes.

Permselectivity is essentially the same for all strong electrolytes with the exception of free H^+ and OH^- ions. In other words, the anion-selective membrane will not exclude H ion efficiently, and the cation-selective membrane is not efficient at excluding OH ions.

The mechanism of electrical conductance in a membrane is similar to conductance in an electrolyte solution. The conductivity depends on the presence of mobile ions to carry the electric current; therefore, ion exchange membranes are more efficient when used to treat solutions which are high in concentration. Less electrical energy is lost to resistive heating.

Chapter 3

Physical Properties of Ion Exchange Products

INTRODUCTION

Since ion exchange materials are in a sense solid electrolytes, it is almost impossible to separate the discussions of physical and chemical properties. There are, however, some physical phenomena which are directly related to the divinylbenzene styrene copolymer structure. These are:

- Density
- Hydration and Swelling
- Resistance to Osmotic Shock
- Diffusion
- Relative Porosity

In addition, the pressure drop, while not dependent on copolymer structure, is a physical property related to the particle size distribution. Today's commercially available ion exchange products are bead-form materials manufactured by suspension polymerization. The discussions which follow are limited to these products.

CROSS-LINKING

Polymerization of styrene by itself yields long linear chains, each containing a large number of styrene monomer units. With no additional treatment, the polymer formed is a solid; however, upon sulfonation, we find that the final product is actually soluble—the solubility being dependent on the degree of sulfonation. This phenomena has been applied industrially, and there are several so-called "liquid ion exchangers" available commercially.

Solubility, however, is not practical when treating water supplies. The use of divinylbenzene as a copolymer in the suspension polymerization provides insoluble finished ion exchange products. The amount of divinylbenzene used can be varied over a wide range. The ion exchange products available today are limited to the range of 2–16 wt% divinylbenzene. Below 2 wt%, the finished ion exchange material lacks the mechanical strength to resist the volume changes which occur during normal operation. Above 16 wt%, the polymer structure resists swelling, so that production of a finished ion exchanger becomes difficult and costly.

DENSITY

For a given ion exchanger type in a known ionic form, the density is directly related to the degree of cross-linking. This relationship for various ion exchangers is given in Tables 3-1 and 3-2. It is also illustrated in Figures 3-1 through 3-4. Manufacturers of ion exchange products will often use several different density values in their literature. The terms used most often—density, absolute density, and shipping weight—are defined in Appendix C (Glossary of Ion Exchange Terms).

SWELLING

The active groups attached to the polymer matrix hydrate when immersed in water. Therefore, the ion exchanger imbibes water when it goes from the dry to the wet state. The amount of water imbibed will be dependent on the nature of the attached active group, the ion exchanged on the group, and the degree of cross-linking. As the cross-linking is increased, the amount of water that the ion exchanger retains will decrease. Therefore, low cross-linked ion exchangers will swell to a greater extent than high cross-linked resins in the same ionic form. As an analogy, one can view the ion exchanger as if it contained springs which expand to accommodate the specific amount of hydration required by the

Table 3-1. Effect of Cross-Linking on the Density of a Strong Acid Cation Exchanger

Cross-Linking (wt%) Divinylbenzene	B & D Density (g/L)	
	Sodium Form	Hydrogen Form
2	120	90
4	260	220
6	360	320
8	420	385
10	460	430
12	490	460
14	510	480
16	520	500

Table 3-2. Typical Hydration Values of Some Basic Ion Exchange Materials

Description of Ion Exchanger	Structure	Ionic Form	Hydration (mmol water/meq)
Strong acid cation (8%XL)	Gel	Na	9.9–11.4
		H	12–15
Strong acid cation (12%XL)	Gel	Na	7.6–9.3
Strong acid cation	Porous	Na	9.3–12
Weak acid cation		H	5.7–7.3
Strong base anion type I (4%XL)	Gel	Cl	12–15
Strong base anion type I (2%XL)	Gel	Cl	13–20
Strong base anion type II (4%XL)	Gel	Cl	9.7–13
Strong base anion type I	Porous	Cl	18–25
Weak base anion	Gel	Free base	18

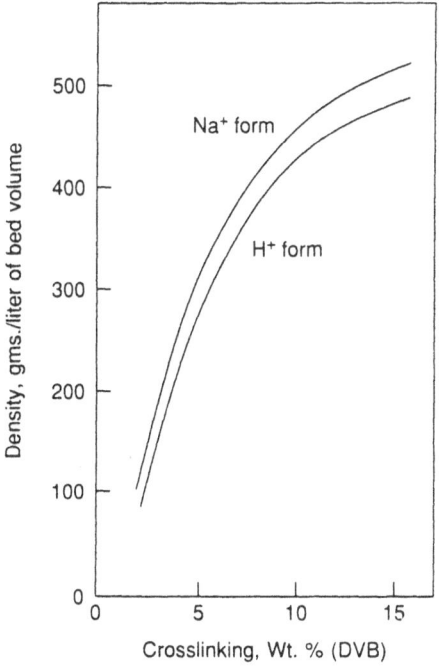

Figure 3-1. Backwashed and drained density as a function of cross-linking for a strong acid gel-type cation resin.

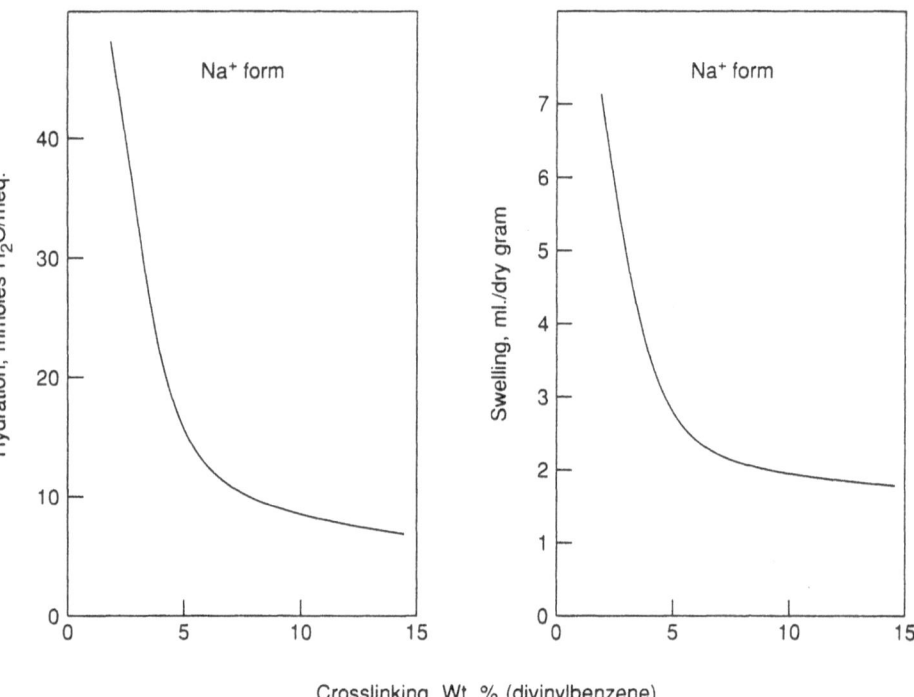

Crosslinking, Wt. % (divinylbenzene)

Figure 3-2. Hydration and swelling of a gel-type strong acid cation exchanger as a function of cross-linking.

physical system. Some swelling (hydration) values for the basic ion exchange materials are given in Tables 3-3 and 3-4.

The information shown in Table 3-3 is typical for most commercially available ion exchange products. The effect of variations in cross-linking on the swelling and hydration of strongly acidic cation exchangers and strongly basic anion exchangers is given in more detail in Tables 3-5 and 3-6, and illustrated in Figure 3-5.

When anion exchange material is placed in a concentrated solution of a salt, there is a competition between the resin and the salt solution for the water. When the solution concentration is high, the salt will desorb water from the resin phase. When this happens, the ion exchanger will shrink. This phenomena occurs during normal applications whenever an ion exchange unit is regenerated. The internal solution phase will cycle from a few % to as high as 10% over a relatively short period of time. As a result, the ion exchange material shrinks and swells many times a year in usual applications, and the ion exchanger will break down. This is often referred to in the water-treatment industry as "osmotic shock" effects.

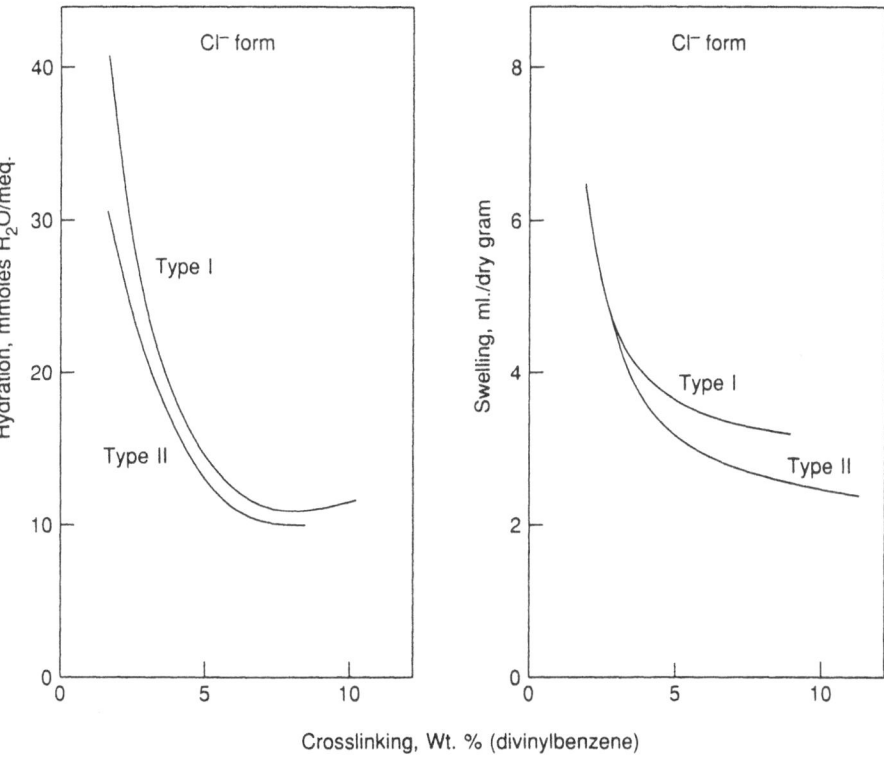

Figure 3-3. Hydration and swelling of gel-type strong base anion exchangers as a function of cross-linking.

For example, a weak acid cation exchanger may swell more than 100% when exhausting from the H form to the Na form. Therefore, the mechanical strength of an ion exchange material plays an important role in the application.

STRENGTH

The strength of a given ion exchange material is, as expected, directly related to the degree of cross-linking. Good ion exchange materials can be characterized as being uniformly spherical, without internal cracks, resistant to mechanical compression, and not brittle. Resistance to "osmotic shock" implies that an ion exchanger is sufficiently flexible to withstand thousands of operating cycles involving shrinking and swelling without developing internal stresses which tend to fracture the beads. Resistance to "attrition" implies that an ion exchanger can withstand the compressive stresses which develop in an operating unit without fracturing. Additionally, in the power-generating industry, it is a

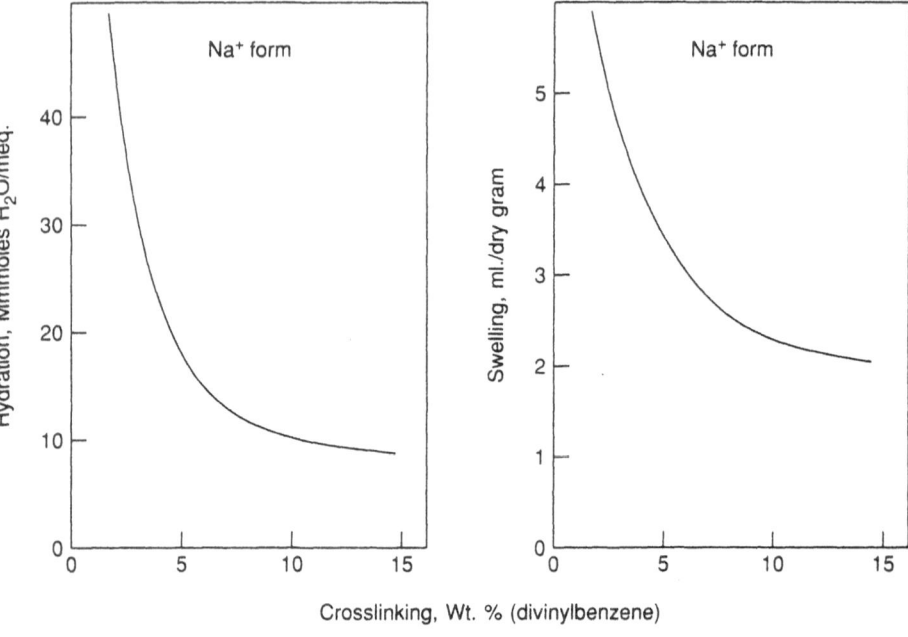

Crosslinking, Wt. % (divinylbenzene)

Figure 3-4. Hydration and swelling of a porous strong acid cation exchanger as a function of cross-linking.

fairly common practice to pump the exhausted ion exchange media to external separation and regeneration equipment.

This practice imposes still other stresses that the ion exchange material must withstand. An extreme example of an application which is very demanding

Table 3.3. The Combined Effects of Cross-Linking and Ionic Form on the Hydration of a Strong Acid Cation Resin

| Cation | Cross-Linking wt% | | |
	4	8 (Hydration, mmol water/meq)	16
Li	23.2	11.7	7.2
H	23.9	11.1	7.6
Na	20.7	10.2	6.3
NH	20.0	9.6	5.9
K	18.9	9.1	5.9
Rb	19.0	8.8	5.9
Cs	19.0	8.8	5.7
Ag	16.1	9.1	5.7

Table 3-4. The Effect of Cross-Linking on Swelling and Hydration of Gel-Type Ion Exchangers

Nominal (wt. % XL)	Strong Acid in Na Form		Type-I Strong Base in Cl Form		Type-II Strong Base in Cl Form	
	Swelling (ml, dry g)	Water (mmol meq)	Swelling (ml, dry g)	Water (mmol meq)	Swelling (ml, dry g)	Water (mmol meq)
2	6.8	45.6	5.9	30.0	5.2	40.0
3	—	—	4.5	22.5	4.5	24.3
4	3.6	20.4	3.9	17.8	3.6	17.9
5	—	—	3.6	14.2	3.2	14.6
6	2.5	12.9	3.5	12.0	3.0	13.1
7	—	—	3.4	11.0	2.8	11.9
8	2.2	9.7	3.2	10.4	2.6	11.5
9	—	—	—	—	2.5	11.4
10	2.0	8.4	—	—	2.5	11.8
12	1.9	8.0	—	—	—	—
14	1.8	7.0	—	—	—	—
16	1.7	6.6	—	—	—	—

of ion exchanger strength is the continuous ion exchange process. Ion exchange materials used in these systems are required to be resistent to:

- Mechanical stress from almost continuous movement.
- Osmotic shock from regenerant concentrations which are often above standard practices.
- Compressive stress which develops because the resin is confined in a restricted volume when it undergoes swelling.

Table 3-5. The Effect of Cross-Linking on Swelling and Hydration of Porous-Type Ion Exchangers

Nominal (wt. % XL)	Strong Acid in Na Form		Type-I Strong Base in Cl Form		Base in Cl Form	
	Swelling (ml, dry g)	Water (mmol meq)	Swelling (ml, dry g)	Water (mmol meq)	Swelling (ml, dry g)	Water (mmol meq)
2	5.7	47.2	7.9	36.9	4.8	33.4
3	—	—	4.9	25.5	3.9	24.1
4	3.9	23.1	3.8	19.7	3.4	19.9
5	—	—	3.4	16.2	3.1	17.1
6	3.1	15.2	3.1	14.2	2.8	14.7
7	—	—	2.9	13.1	2.6	13.3
8	2.5	11.7	2.7	12.3	2.5	11.8
9	—	—	2.6	11.6	—	—
10	2.3	10.0	2.5	12.1	—	—
12	2.1	8.9	—	—	—	—
14	2.0	8.5	—	—	—	—

Table 3.6.   The Effect of Cross-Linking on the Diffusion Rates in a
Gel-Type Strong Acid Cation Exchanger

Ion	Valence	Cross-Linking (wt%)	Solid Diffusion Rate (cm/sec at)	
			0.3 C	25 C
Na	+1	4	6.7E-7	1.4E-6
		8.6	3.5E-7	9.4E-7
		10	1.2E-7	2.9E-7
		16	6.6E-8	2.4E-7
		24	2.7E-8	1.0E-7
Ag	+1	8.6	2.6E-7	6.4E-7
		16	1.0E-7	2.8E-7
		24	3.8E-8	1.1E-7
Cs	+1	8.6	6.6E-7	1.4E-6
		24	3.3E-8	7.3E-8
Zn	+2	2	3.0E-7	7.4E-7
		8	2.1E-8	6.3E-8
		10	8.8E-9	2.9E-8
		16	3.1E-9	1.2E-8
		24	5.5E-10	2.6E-9
Sr	+2	4	9.7E-8	2.3E-7
		10	9.6E-9	3.4E-8
		16	5.6E-10	3.0E-9
Y	+3	2	4.1E-8	1.1E-7
		4	2.7E-8	7.5E-8
		10	1.0E-9	3.2E-9
		24	5.9E-11	2.2E-10
La	+3	2	5.2E-8	1.3E-7
		4	3.0E-8	6.9E-8
		8	3.0E-9	9.2E-9
		16	1.5E-10	5.0E-9
Th	+4	10	6.4E-11	2.2E-10

As indicated by the above discussion, each application must be reviewed for the amount of stress imposed on the ion exchange resin being employed. This is important since the ion exchanger represents a large capital cost in any water-treatment system; and, properly applied, they are expected to give many years of effective service.

DIFFUSION

An ion, in order to exchange with a counter-ion at an active site, must diffuse through the boundary layer and the solid resin phase to a point where it can react with the attached counter-ion. In addition, the counter-ion, now free, must

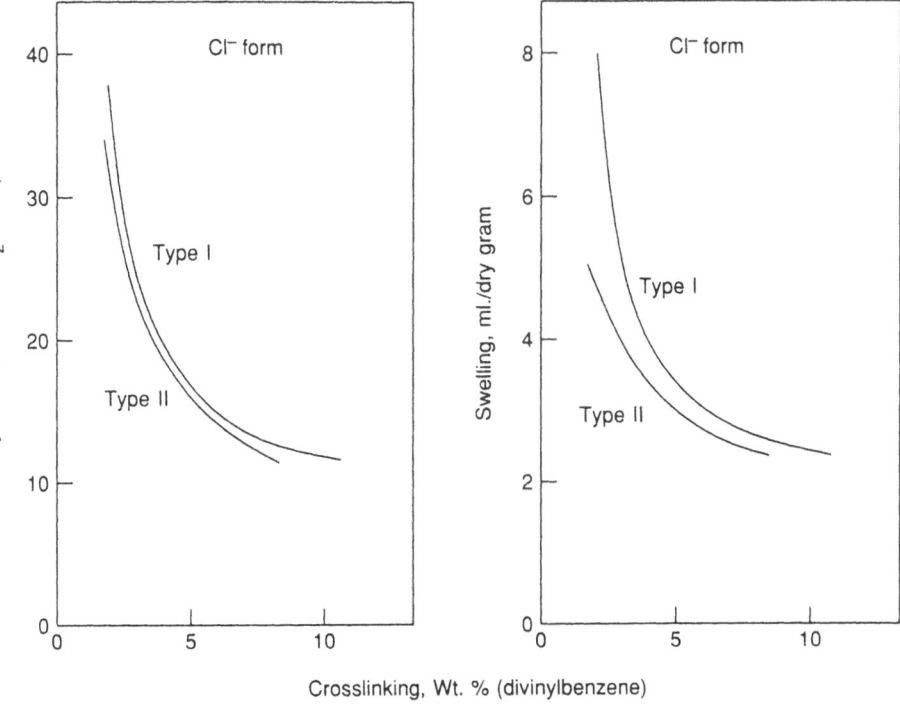

Figure 3-5. Hydration and swelling of porous strong base anion exchangers as a function of cross-linking.

diffuse in the opposite direction to the external solution phase to complete the reaction. This diffusion takes time and is effected by the bead size and degree of cross-linking. The relative effect of cross-linking on the solid phase diffusion rates of several ionic species is shown in Table 3-7.

Table 3-7. The Effect of Cross-Linking on the Diffusion Rate of Bromine in a Type-II Strong Base Anion Exchanger (Gel Type)

Cross-Linking (wt%)	Solid Diffusion Rate (cm/sec at)	
	0.3 C	25 C
1	4.4E-7	9.1E-7
2	3.0E-7	6.4E-7
3	2.0E-7	4.5E-7
6	1.5E-7	3.9E-7
8	6.1E-8	2.0E-7
16	6.0E-8	2.6E-7

POROSITY

Most common ions are relatively small (< 10 Å). However, some natural organic acids can have very high moleular weights and will be found to have sizes in excess of 100 Å. These large organic acids will exchange on anion resins, and, because of their size, are often very difficult to remove by normal regeneration processes. Therefore, porous ion exchange materials have been developed to accommodate these operating conditions. Some of these products have pore diameters of over 100,000 Å. There are currently two methods used for the preparation of porous exchangers:

- In one case, a linear monomer with a controlled molecular weight is added to the styrene-divinylbenzene monomers as they polymerize. When the final ion exchange product is prepared, cation or anion exchanger, the "foreign" polymer is converted to a soluble species which is washed from the ion exchange material leaving the pores. The pores are formed by "extraction" of the soluble organic electrolyte.
- The most common process used is known as *phase separation*. In this case, a solvent is added to the styrene-divinylbenzene monomers. The solvent is selected to be soluble in the monomers and precipitates as the polymerization goes to completion. Pores are left in the finished polymer when the solvent is removed. The pores formed by this process can be varied, and the final products are more porous than those prepared by the "extraction" technique.

Chapter 4
CHEMICAL PROPERTIES OF ION EXCHANGE PRODUCTS

INTRODUCTION

As the term implies, the process of ion exchange is the exchange of ions. In this context it is specifically the exchange of ions between a solid phase, the ion exchange matrix, and a liquid phase—generally, water. While most of the ion exchange processes apply to water treatment, it is not the only medium in which the ion exchange process can be applied. There are numerous applications where ion exchangers have been used to purify or catalyze reactions in mixed and nonaqueous solvents.

Since the ion exchange process involves two different media, a solid and a liquid, it is often referred to as a heterogeneous system. While the general rules of homogeneous equilibria apply, the chemical properties of the ion exchange material have an important impact on the final results. Heterogeneous equilibria can be characterized by the following statements:

- Any increase in the interface area will increase the reaction rate but will not alter the final equilibrium concentrations or conditions.
- At a constant temperature, each system has a unique equilibrium state, which can be defined by a unique equilibrium constant.
- An increase in temperature will increase the reaction rate, and may alter the equilibrium constant. When this occurs, the change will be in the direction that absorbs heat or energy.

The principal chemical properties of an ion exchange material, which effects their performance and application, are

- Hydration
- Ionization
- Equilibria and Selectivity

These properties are related to the matrix structure and the chemical nature of the active group attached to the polymer matrix. The ion exchanger is considered as a soluble electrolyte restrained by the cross-linking used in the preparation of the initial polymer. Each of these attributes will be discussed in some detail in the sections to follow.

HYDRATION

In all ion exchangers the ionization of the attached active group is dependent on the presence of water in the matrix. The amount of water an ion exchange material will imbibe, in turn, is dependent on the cross-linking (i.e., rigidity) of the polymer. This dependence is shown in Table 4-1 and 4-2 for a strong acid cation exchanger, and two strong base anion exchangers. Similar data for a weak acid cation exchanger is given in Table 4-3: however, similar information for weak base anion exchangers is not available at this time.

IONIZATION

In aqueous (water) media, strong acid cation and strong base anion exchange resin are fully hydrated; and the ions associated with the active group are always free to exchange with ions of like charge in the solution being processed. The value at which ionization becomes effective (pK value) in weak acid cation and weak base anion exchangers is different, as shown in Table 4-4.

Table 4-1.   Effect of Cross-Linking on the Water Retention of Gel Ion Exchangers

| Nominal (wt%) | Water Retention, wt% | | |
| | Strong acid cation | Strong base anion (Cl form) | |
	(Na form)	Type I	Type II
2	79	71	73
3	—	61	62
4	64	55	54
5	—	50	49
6	52	45	46
7	—	43	43
8	45	40	42
9	—	—	41
10	41	—	40
12	39	—	—
14	36	—	—
16	33	—	—
18	31	—	—
22	28	—	—

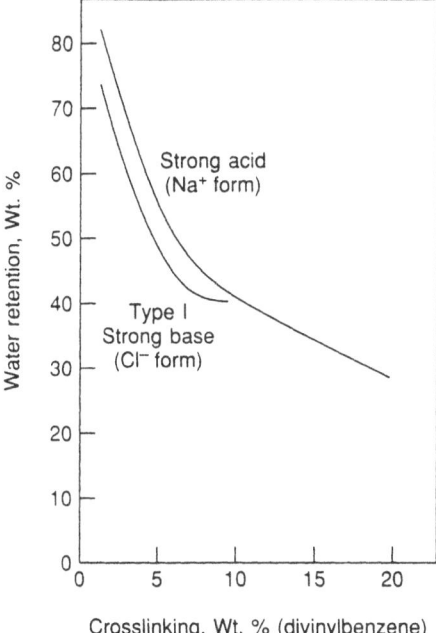

Crosslinking, Wt. % (divinylbenzene)

Figure 4-1. Water retention of gel ion exchangers as a function of cross-linking.

Table 4-2. Effect of Cross-Linking on the Water Retention of Porous Ion Exchangers

	Water Retention, wt%		
Nominal (wt%)	Strong acid cation (Na form)	Strong base anion (Cl form)	
		Type I	Type II
2	80	73	68
3	—	66	60
4	66	60	55
5	—	55	51
6	56	52	46
7	—	49	43
8	50	47	40
9	—	45	—
10	45	44	—
12	42	—	—
14	40	—	—

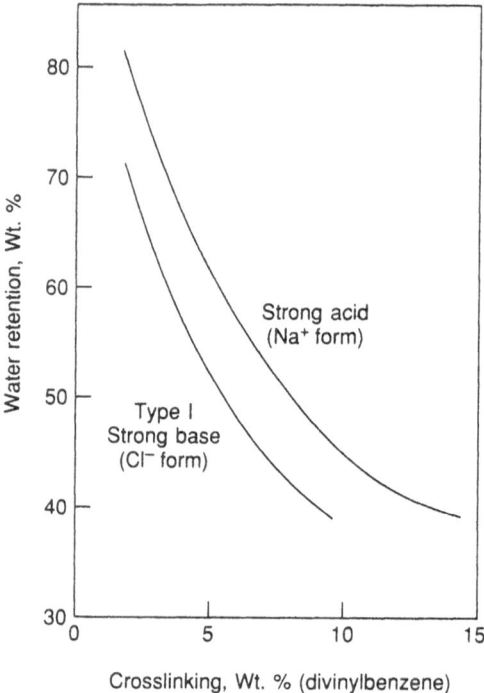

Figure 4-2. Water retention of porous ion exchangers as a function of cross-linking.

The information shown in Table 4-4, indicates that, in general terms:

- The strong acid cation exchanger will be effectively ionized at any pH greater than 1.
- The strong base anion exchanger will be effectively ionized at any pH lower than 13.

Table 4-3.   Effect of Cross-Linking on the Water Retention of a Weak Base Anion Exchanger

Cross-linking (wt%)	Water Retention (wt%)
2	48
4	40
6	31
8	24
10	17
14	7

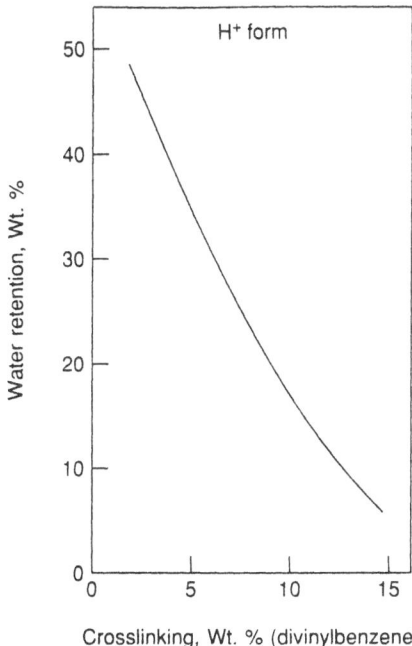

Figure 4-3.   Water retention of a weak acid cation exchanger as a function of cross-linking.

In contrast:

- The weak acid cation exchanger does not become effectively ionized, depending on it's preparation until the pH is above the range shown.
- The weak base anion exchanger in the same way is not effectively ionized unless the pH is lower than the range shown in Table 4-4.

Table 4-4.   Ionization and pK Values of Ion Exchange Materials

Ion exchange material	Active group	Approximate pK value
Strong acid cation	$R-SO_3$	<1
Weak acid cation	$R-COH$	4–6
Strong base anion		
Type I	$R-N(CH_3)_3$	>13
	C_2H_5OH	
Type II	$R-N<$	>13
	$(CH_3)_2$	
Weak base anion	$R-N(CH_3)_2$	7–9

- The salt forms of weakly functional ion exchangers will undergo hydrolysis in water, as shown by the following equations.

Weak acid cation resin

$$R-COONa + H_2O \rightleftharpoons R-COOH + NaOH \qquad (4\text{-}1)$$

Weak base anion resin

$$R-N(CH_3)_2 - HCL + H_2 \rightleftharpoons R-N(CH_3)_2 + HCl \qquad (4\text{-}2)$$

Therefore, when a sufficient volume of water is passed through an exhausted column or operating unit the ion exchanger can be converted to its original regenerated form. The following sections deal with the characteristics of the weakly functional ion exchangers in some detail.

WEAK BASIC ANION EXCHANGERS

Weak base ion exchange products are those materials which have primary and secondary amines as the active group. These products are manufactured by some suppliers with mixed amine groups. Those products with tertiary amines will exhibit some salt-splitting capacity and as such are often referred to as intermediate base anion exchangers. The active groups of a true weak base anion exchanger will not dissociate in waters where the pH is lower than their characteristic pK value. As a result, weak base anion exchangers are very effective for the removal of free acids, such as NSO_4, HCl, and HNO_3 from water. In addition, when the solution pH is below the pK value, a weak base anion exchanger will react with slightly acidic salts, such as ammonium chloride. In the absence of any salt-splitting ability, neutral salts, such as, $NaCl$, Na_2SO_4, and $NaNO_3$, will not be exchanged by weakly basic anion exchange materials.

Weak base anion exchange materials are easily regenerated to their free base form. Regeneration can be accomplished by using $NaOH$, HN_4OH, or Na_2CO_3, for example. Usually complete regeneration can be achieved with about 10% of the stoichiometric amount defined by the operating capacity in a given application. The operating capacity is related to the concentrations of exchangeable acids in the water being processed. One disadvantage that weak base ion exchangers have is their sensitivity to process flow rate as shown in Table 4-5.

WEAK ACID CATION EXCHANGERS

The chemistry of weak acid cation exchangers is quite different and somewhat unique. The unique properties of these materials are specific selectivity and increased operating efficiency. To take advantage of these unique properties, it is necessary to have some understanding of the behavior of weak acid cation exchangers in water and waste-water applications.

Table 4-5.   Weak Base Anion
Exchange Capacity as a Function
of Operating Flowrate

Operating flowrate (g/f^3)	% capacity at 2 g/f^3
1	103
2	100
4	93
11	68[a]

[a]Extrapolated value.

Weak acid cation exchangers are essentially unionized when the solution pH is below the characteristic pK value of the product being used. Weak acid cation exchangers usually exhibit ion exchange behavior above their pK value, and full capacities are only realized when the water pH is above eight. Most available weak acid cation exchangers have a capacity of 4 meq/ml or more for the free acid form. Therefore, useful operating capacities can be realized when only a portion of the total capacity is available in an application.

The relationship between the degree of ionization and solution pH is critical to the application of these products. The ionization of the hydrogen-form resin can be written as follows:

$$R-COOH \rightleftharpoons R\text{-}COO^- + H^+ \tag{4-3}$$

Less than 1% of the carboxylic groups are ionized when the hydrogen-form material is in contact with pure water. In order to understand the behavior of a weak acid cation exchanger, the following specific topics will be reviewed in detail.

• Reaction with a base
• Reaction with a basic salt
• Reaction with a neutral salt
• Effects of anions and cations
• Salt–salt exchange
• Capacity and regeneration

REACTION WITH A BASE

When hydrogen ions are removed from the right-hand side of equation (4-3), the resin will be converted to the more highly ionized form:

$$R-COOH + Na^+ + OH^- \rightleftharpoons R-COONa + H_2O \tag{4-4}$$

Actually, sodium hydroxide reacts with the weak acid groups to form a highly ionized resin salt; similar to the reaction of sodium hydroxide with acetic acid to form the ionized sodium acetate salt. The conversion of the resin to the salt form is more complex because the reactive group is permanently attached to the polymer matrix. Electrostatic interactions within the polymer matrix increases the difficulty for the ionization to go to completion. A pH of 8-10 is required to convert the hydrogen form of the weak acid ion exchanger to the sodium form. In addition, for site neutralization to take place, the sodium ion must diffuse to the active site, and the free hydrogen ion must diffuse out to the solution so that electroneutrality is maintained. While diffusion is common to all ion exchange processes, it is especially critical with weak acid cation exchange materials, since this diffusion will determine the operating flowrate for a particular application.

REACTION WITH A BASIC SALT

The reaction of a weak acid cation exchanger with a solution of sodium bicarbonate can be expressed as follows:

$$R-COOH + Na^+ + HCO_3^- \rightleftharpoons R-COONa + H_2CO_3 \qquad (4\text{-}5)$$

On inspection, this looks like equation (4-4), since it yields slightly ionized products in solution; however, the ionization product of the carbonic acid formed is on the same order of magnitude as the degree of ionization of the hydrogen form of the weakly acidic cation exchanger (see equation 4-3). Therefore, the available hydrogen ion from the carbonic acid limits the reaction. As a result, the reaction (4-4), is more characteristic of a mass-action-controlled process than of a neutralization. This is an important concept in understanding the performance of weak acid cation exchange materials.

REACTION WITH A NEUTRAL SALT

Weak acid cation exchangers react with neutral salts to a very minor degree.

$$R-COOH + Na^+ + Cl^- \rightleftharpoons R-COONa + H^+ + Cl^- \qquad (4\text{-}6)$$

This is because one of the products of the reaction is a highly ionized acid. The solution pH will therefore be low, and as a result the ionization of the weak acid group on the polymer matrix will be depressed and driven to the left (see equation 4-4). While the reaction is minor, it can be observed. When a solution of sodium chloride is passed through a weak acid cation exchanger, in the hydrogen form, the effluent pH will be lower than the influent. However, this is a "mass action" effect, and quite unlike the salt-splitting reaction we observe with the strong acid and strong base ion exchanger materials. The degree of pH-lowering is dependent on the influent sodium chloride concentration and is, therefore, consistent with the "mass action" concept.

EFFECT OF ANIONS AND CATIONS

The nature of the cation in the solution being processed has little effect when the weak acid cation exchanger is used for neutralization, this is also true for strong acid cation exchange materials; since the reactions are very similar:

$$R-COOH + Na + OH \rightleftharpoons R-COONa + H_2O \qquad (4\text{-}7)$$

$$R-SO_3 + Na + OH \rightleftharpoons R-SO_3Na + H_2O \qquad (4\text{-}8)$$

In contrast, when a weak acid cation exchanger reaction is subject to "mass action" behavior (see equation 4-4), the effect of a particular ion may be controlling in some cases. This is important for dealkalization reactions, for example:

$$R-COOH + NaHCO_3 \rightleftharpoons R-COONa + H_2CO_3 \qquad (4\text{-}9)$$

$$2R-COOH + Ca + 2HCO_3 \rightleftharpoons (R-COO)_2 - Ca + 2H_2CO_3 \quad (4\text{-}10)$$

The difference between equations (4-9) and (4-10) is due to the difference in the properties of the resin salts formed. The "mass action" selectivity of the weak acid cation exchanger for calcuim over sodium ion is on the order of 50. Therefore, as expected, the weak acid cation exchanger prefers to form the calcium salt rather than the sodium salt, and the reaction is driven to the left.

It is not possible, at present, to put actual numbers on the equilibria for reactions (4-9) and (4-10). However, actual column experiments have shown that the equilibrium is unfavorable for reaction (4-10) and favorable for (4-9). As expected then, the performance of a weak acid cation exchange material in dealkalization is very dependent on influent water composition and pH (the ratio of total hardness to alkalinity becomes a very important parameter). In the same way, the equilibrium, and the efficiency of a column operation will be effected by the anion in the influent water being processed. The degree of ionization of the acid form will influence the degree of reaction to the left in equation (4-9). Therefore, water treatment by a weak acid cation exchange material is more efficient when processing a water-containing carbonate ion than one containing only bicarbonate ions.

SALT–SALT EXCHANGE

Salt forms of weak acid cation exchangers will react with salts in solution, as long as the influent pH remains high enough to ensure that the carboxylic acid group is in the ionized form. Since weak acid cation exchange materials exhibit higher selectivities for divalent ions, they offer an interesting potential for metal separations at operating conditions which would make strong acid cation exchangers relatively inefficient.

CAPACITY AND REGENERATION

The relationship between operating capacity and regenerant dosage is very different for a weak acid and a strong acid cation exchanger. It is usually impractical to fully regenerate a strong acid cation exchange material, except in special cases when the high cost can be justified.

In contrast, a weak acid cation exchanger can be regenerated to the hydrogen form with an amount of regenerant that is essentially equal to the amount of ions exchanged on the column during the operating cycle. The overall utilization may be low, for a given application, but the regeneration efficiency is always high with respect to the ions removed by the treatment process. This is never possible with a strong acid cation exchange material.

Normal practice is to set the regenerant dosage at 105% of the operating capacity for the application. The amounts of regenerant required by any water-treatment application can be calculated by using one of the following mathematical expressions:

$$\frac{\text{g (100\% HCl)}}{\text{L}} = \text{oper. capac. } \frac{\text{equiv.}}{\text{L}} \times 36.5 \times 1.05$$

$$\frac{\text{lb (100\% HCl)}}{\text{f}^3} = \text{oper. capac. } \frac{\text{kg (as CaCO}_3)}{\text{f}^3} \times 0.104 \times 1.05$$

$$\frac{\text{g (100\% H}_2\text{SO}_4)}{\text{L}} = \text{oper. capac. } \frac{\text{equiv.}}{\text{L}} \times 49 \times 1.05$$

$$\frac{\text{lb (100\% H}_2\text{SO}_4)}{\text{f}^3} = \text{oper. capac. } \frac{\text{kg (as CaCO}_3)}{\text{f}^3} \times 0.144 \times 1.05$$

Any strong mineral acid, and many weak acids, can be used to regenerate a weak acid cation exchange material. Note, however, that the weak acid cation exchanger will remove calcium ions from most water supplies in preference to sodium ions. Therefore, when regenerating with sulfuric acid, care must be taken to keep the concentration well below the solubility limit of calcium sulfate. The concentration usually used in the water-treatment industry is in the range of 0.5–0.8 wt%. Fortunately, the weak acid cation exchange material will regenerate efficiently even at this low concentration.

EQUILIBRIA AND SELECTIVITY

It is understood that ion exchange reaction are reversible equilibria. During regeneration, we employ an excessive amount of electrolyte to drive the reaction in a preferred direction. This is shown in the following equation, which represents the removal of ion B from a process stream.

$$R - A + B \underset{\text{regeneration}}{\overset{\text{process}}{\rightleftharpoons}} R - B + A \tag{4-11}$$

However, when a limited amount of electrolyte is available, the equilibrium established depends on the concentration of ions A and B in the process water, and also on the selectivity of the ion exchange material being used.

The selectivity coefficient, K_B^A, for the reaction, (4-11), is defined as follows:

$$K_B^A = \frac{(meq/g, B_r)}{(meq/g, A_r)} \times \frac{(meq/ml, A_s)}{(meq/ml, B_s)} = \frac{(B_r)}{(A_r)} \times \frac{(A_s)}{(B_s)} \qquad (4-12)$$

where the subscript, r, refers to the solid ion exchanger phase, and the subscript, s, refers to the solution phase. This selectivity coefficient is not quite the same as the equilibrium constant, which is related to activities rather than the measured concentrations. To obtain a thermodynamically correct selectivity coefficient, the result from equation (4-12) must be multiplied by the activity coefficients of the ions of interest in the resin and water phases.

The existing theories of ion exchange are for the most part a combination of two older theories, each contributing to the overall description of ion exchange equilibria. In the first theory a Donnan potential was proposed between each ion exchange particle and the surrounding solution. As developed by Bauman and Eichorn, the ion exchanger is regarded as a soluble electrolyte restrained by the polymer cross-linking. At low concentrations, the counter-ions are assumed to be essentially excluded from the resin phase, and the ionic activities product of the electrolyte is the same for both phases.

The second, and quite different, theory (as introduced by Gregor) defines the exchange selectivities by incorporating the polymer swelling or hydration, which is associated with the presence of different ions. Since energy is required to swell the resin, such as stretching a spring, the ion exchanger will be selective for the hydrated ion, which occupies the smallest volume. As a result the theory is essentially Hooke's Law.

At this time these two basic concepts have not been discarded, both are considered to contribute to the description of observed ion exchange phenomena. A refined approach, validated by Glueckauf, included both theories as follows:

$$\ln K_B^A = \frac{P(V_a - V_b)}{RT} + \ln (B_r/A_r)$$

where P = the internal swelling or osmotic pressure
V_a = molar volume of the A ion in the resin phase
V_b = molar volume of the B ion in the resin phase
B_r = moles of ion B in resin phase
A_r = moles of ion A in resin phase
RT = 0.0821 × absolute temperature

Table 4-6. Selectivity Coefficients of Cations on a Strong Acid Cation Exchanger[a]

Cross-linking, wt%	4	8	12	16
Monovalent ions				
H	1.0	1.0	1.0	1.0
Li	0.9	0.85	0.81	0.74
Na	1.3	1.5	1.7	1.9
NH	1.6	1.95	2.3	2.5
K	1.75	2.5	3.5	4.5
Rb	1.9	2.6	3.1	3.4
Cs	2.0	2.7	3.2	3.45
Cu	3.2	5.3	9.5	14.5
Ag	6.0	7.6	12.0	17.0
Divalent ions				
Mn	2.2	2.35	2.5	2.7
Mg	2.4	2.5	2.6	2.8
Fe	2.4	2.55	2.7	2.9
Zn	2.6	2.7	2.8	3.0
Co	2.65	2.8	2.9	3.05
Cd	2.8	2.95	3.3	3.95
Ni	2.85	3.0	3.4	4.15
Ca	3.4	3.9	4.6	5.8
Sr	3.85	4.95	6.25	8.1
Hg	5.1	7.2	9.7	14.0
Pb	5.4	7.5	10.1	14.5
Ba	6.15	18.7	11.6	16.5

[a]The choice of hydrogen as the reference ion is arbitrary.

Table 4-7. Selectivity Coefficients of Monovalent Anions on Strong Base Anion Exchangers

Anion	Type I	Type II
Hydroxide	1.0	1.0
Benzenesulfonate	> 500	75
Salicylate	450	65
Citrate	220	23
Iodide	175	17
Phenate	110	27
Bisulfate	85	15
Chlorate	74	12
Nitrate	65	8
Bromide	50	6
Cyanide	28	3
Bisulfite	27	3
Bromate	27	3

Table 4-7.   (*Continued*)

Anion	Type I	Type II
Nitrate	24	3
Chloride	22	2.3
Bicarbonate	6.0	1.2
Iodate	5.5	0.5
Formate	4.6	0.5
Acetate	3.2	0.5
Propionate	2.6	0.3
Fluoride	1.6	0.3

It is well-known that many other factors effect ion exchange equilibria such as:

- Ion-pair formation
- Salting-out effects
- Repulsion due to like charges
- Degree of site ionization

The real nature of ion exchange equilibria is both complicated and varied, but the end result is a broadening of the utility of the ion exchange unit process. This versatility is evident from the selectivity coefficients (equilibrium constants) of various ions for both strong acid cation exchangers (Table 4-6) and strong base anion exchangers (Table 4-7). Similar data related to weakly acidic and weakly basic ion exchange materials are not available in the literature of water-treatment technology.

Chapter 5
CHROMATOGRAPHY

The process of chromatography was first developed by M. Twsett in 1906. The initial technique, based on absorption and not ion exchange, remained dormant for 25 years. In 1931, chromatography was still essentially an absorption process which relied on finely divided solids (i.e., aluminum oxide, silica, etc.) to separate complex natural organic compounds. The fact that many of these compounds were colored, is the reason for the name of the process—namely, chromatography. This remained the primary application of chromatography until the early 1940s.

The availability of commercial bead-form ion exchange products with uniform particle size, both strong and weak activities, sparked new interest; and the technology known as ion exchange chromatography was born.

The process was first used to separate the "rare earths." These interesting elements are closely related chemically. They differ in mass by less than 1%, and are very difficult to separate as pure compounds by conventional wet chemical techniques; however, cation exchange chromatography worked! Chromatographic techniques separated these "rare earths" into pure fractions, and shed light on the chemistry of numerous "fission products" produced during the Manhattan Project.

Ion exchange chromatography has been used to analyze trace concentrations of cations and anions in polluted water supplies, so that pollution can be traced to a specific source. Since very small quantities of anions and cations can be concentrated on an ion exchanger and separated into specific species, the process has become a very useful analytical tool. The analysis of trace elements in "ultra pure" water-treatment systems is routinely accomplished down to a few "parts per billion" (ppb).

To this date, ion exchange chromatography is generally considered to be an analytical tool. There are very few production-scale applications of the process.

Although the separation of "rare earths" has been mentioned, more recently ion exchange chromatography has been applied to the separation of hexose, glucose, and fructose sugars, which are the principle components of invert sugar. This process was developed to find a natural sweetener with low calorie content. Fructose was found to have the highest sweetening level, as less is required, lower calorie intake is achieved. Understandably then, a review of the basic principles of ion exchange chromatography is of some importance.

ION EXCHANGE CHROMATOGRAPHY

As indicated earlier in this chapter, chromatography was first applied to the separation of natural organic compounds using solid inorganic absorbents, such as alumina, clays, activated carbon, and silica. Later, ion exchange chromatography became the technique for delicate separations in analytical and preparative chemistry.

Conventional and ion exchange chromatography are similar in many respects. The mechanisms are, however, very different:

- Conventional chromatography relies basically on sorption and desorption for the separation of essentially nonionic compounds by a solid sorbent.
- Ion exchange chromatography relies on the stoichiometric exchange of counter-ions between the exchanger and the mixture of ions being separated.

In actual practice, a black-and-white distinction cannot be made, since many inorganic sorbents exhibit some ion exchange properties, and ion exchange processes are known to be accompanied by both sorption and hydrolysis.

While chromatography columns may be operated in different ways, there are three basic techniques:

- Displacement Development
- Elution Development
- Frontal Analysis

Each of these will be reviewed in some detail in the following sections.

DISPLACEMENT DEVELOPMENT

Consider a solution containing three counter-ions, A, B, and D, which need to be separated by displacement development. The ion exchanger in this case must be chosen so that the relative selectivities are $B < C < D$. In addition, the ion exchanger must also be converted to the A form so that $A < B < C < D$. Development separation is accomplished with another counter-ion, which the ion exchanger of choice prefers, so that finally $A < B < C < D < E$.

During the development by a solution containing a compound EY, where Y is a common co-ion, A, B, C, and D are displaced down the column and self-sharpening boundaries are formed as shown in Figure 5-1. As the elution continues, the individual bands follow each other down the column without

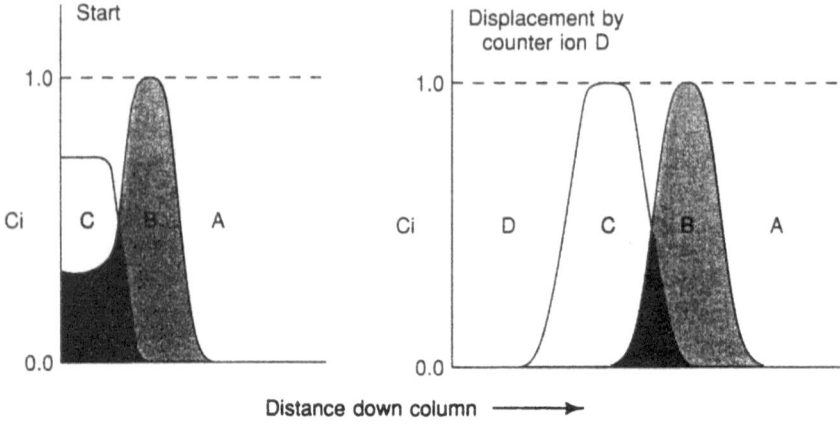

Figure 5-1. Displacement development.

intervals at equal rates. The rate of migration down the column is dependent on the flowrate, feed concentration, and volume capacity of the ion exchanger being used. It is completely independent of the counter-ion concentrations. In the effluent, the bands will appear in the order of their selectivity—A, B, C, D, and E. Therefore, displacement chromatography will yield pure fractions of the compounds BY, CY, and DY, with small overlaps which can often be recycled.

This technique is primarily suited to preparative separations, and is used extensively in the pharmaceutical industry for the separation of pure protein fractions.

The development of sharp boundaries and good separation is enhanced by:

- High Selectivity
- High Volume Capacity
- Small Particle Size
- Low Flowrate
- Low Feed Concentration
- Elevated Temperatures

ELUTION DEVELOPMENT

Again, consider a solution containing the same counter-ions B, C, and D to be separated. The same ion exchanger can be used regenerated in form A. Now the selectivity is A < B < C < D. A small quantity of the mixed solution is introduced to the top of the ion exchange column. The total equivalents of the mixed sample added needs to be less than 5% of the column capacity in order to avoid "overloading." Overloading will result in interferences of the peaks developed and will severely reduce the separation efficiency.

The mixture (B, C, D) is developed by eluting with a solution of AY, where A is the counter-ion with the lowest affinity for the ion exchanger sites among the other ions in the mixture to be separated. Therefore, the counter-ion A will bypass the other ions which are more strongly held by the ion exchanger. The ions B, C, and D will migrate down the column in the presence of A at different rates. Therefore, separation of B, C, and D occurs as they grow further and further apart; this is illustrated by Figure 5-2.

The bands have nonsharpening boundaries and will flatten out on their way down the column. In the effluent, the individual species B, C, and D will appear at low concentrations, and complete resolution of the mixture can be

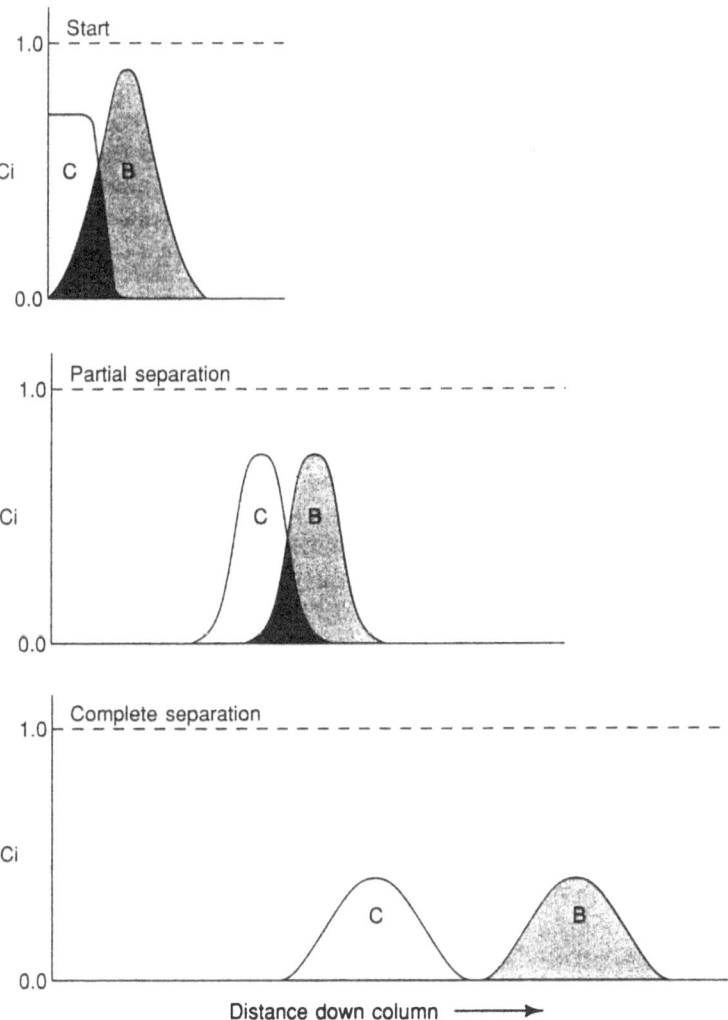

Figure 5-2. Elution development.

obtained. This technique is similar to gas chromatography and is well-suited to quantitative analysis of elements at concentrations of a few parts per billion (ppb).

The sharpness of the separated peaks depends on how well the local equilibrium in the column layers is approached. For example, sharp peaks are favored by:

- High Ion Exchange Rates
- Low Migration Rates
- Small Particle Size
- Low Flow Rates
- Elevated Temperatures

An increase in column length will improve separation, but, may at the same time cause excessive flattening of the eluted peaks. As much as possible, operating conditions should be chosen so that no overlap occurs. In addition, unnecessary gaps between the peaks should be avoided to save time and reagents.

In special cases, stepwise or continuous change of the eluent composition, gradient elution, will provide distinct advantages. The gradient elution technique has been used for the separation of amino acids on a strong acid cation exchanger.

FRONTAL ANALYSIS

Frontal analysis is different than the two methods already described, since in this application the mixture B, C, D to be separated is passed continuously through the ion exchange column. In this case, the counter-ion A is less strongly held by the resin being used, and the selectivity series A < B < C < D still applies. With the chromatography column regenerated in the A form, the mixture displaces species A and a self-sharpening boundary forms. As the process continues, the counter-ions appear in the effluent in the reverse order of their selectivity: B will be first, followed by C, and finally by D (as illustrated in Figure 5-3).

With frontal analysis, however, only one counter-ion of the mixture, A, is isolated. The remaining ions of the mixture are recognized by the occurrence of "fronts" in the column effluent. The relative amounts of each component can be calculated only when the selectivity of each is known with respect to the common counter-ion A.

Frontal development is not applicable to quantitative analysis or preparative processes; therefore, it is not used industrially.

COMPLEXING AGENTS

As indicated earlier, chromatography separations are based on the differences in selectivity of the ion exchange being used. As a result, separations will not occur unless the ion exchanger can differentiate between the ions to be sepa-

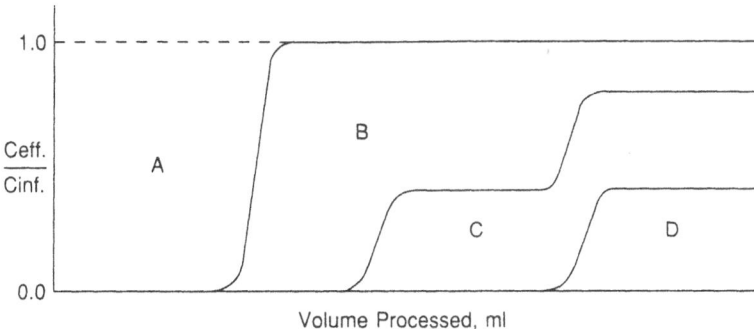

Figure 5-3. Frontal analysis.

rated. Elements with similar chemical characteristics as the "rare earths" are very difficult to separate by conventional chromatography. Complexing agents, such as:

- Citrate
- Lactate
- Ethylenediaminetetraacetate
- Nitrolotriacetate
- α-Hydroxyisobutyrate

have been used to modify the selectivity of closely related chemical species with respect to the ion exchange material, so that good separations are possible.

The use of complexing agents made possible the separation of the "rare earths" in kilogram quantities.

The effect of a complexing agent on the distribution of an ion, B, at equilibrium is influenced by the following two competing reactions:

$$B + A - R ===== B - R + A \qquad (5\text{-}1)$$

$$B + Y^n ===== B - Y^{(n-1)} \qquad (5\text{-}2)$$

Equation (5-1) is recognized as an ion exchange equilibrium and can be characterized by a suitable equilibrium constant. Equation (5-2) is characterized by a stability constant for the complex formed.

For example, when similar chemical ions B and C are separated by chromatography in the presence of a complexing agent, the ion forming the strongest complex (i.e., higher stability constant) will be eluted first, often without C ions in the effluent. The C ions, on the ion exchanger, can now be eluted as a pure fraction by continuing the elution without the complexing agent.

HEIGHT EQUIVALENT OF A THEORETICAL PLATE (HETP)

A knowledge of the HETP is helpful in any ion exchange application; however, it is particularly useful when sensitive chromatographic separations are being done.

In elution development, the bands of ions move independently of one another in the linear range of the distribution isotherm. Therefore, it is possible to estimate the rate of movement down the column as follows:

$$\frac{cm}{ml} = \frac{1}{(D + a)} \tag{5-3}$$

where D = the distribution coefficient
a = the void fraction of the column (~ 0.4)

The elution volume required for the appearance of the concentration maximum is:

$$v_{max} = X(D + a) \tag{5-4}$$

and, X, is the total volume of the chromatographic column. It follows that the maximum concentration, C_{max} can be estimated using:

$$c_{max} = \frac{m}{v_{max}} \times \sqrt{\frac{N}{2\pi}}$$

where m is the total amount of ion expressed as moles, and N is the number of theoretical exchange plates for the column being used. Therefore, when C_{maxP} and m are known from analytical data, and v_{max} is measured, the number of theoretical plates, N, is easily estimated with equation (5-4). So that HETP = L/N, where L is the column length measured in cm.

A somewhat simpler method of estimating N is:

$$N = 8\left(\frac{v_{max}}{\beta}\right)^2$$

The value of β is the width of the eluted band when the concentration of the eluted species is equal to $0.368 \times C_{max}$. From the above, it is apparent that the HETP is a very important parameter for ion exchange chromatographic applications.

At the present time, ion exchange chromatography is being applied to:

• Separation of important pharmaceuticals.
• Partial recovery of valuable materials in waste streams.
• Analysis of trace polutants in streams and industrial wastes.
• Analysis of cations and anions down to the ppb concentration.

Chapter 6

ION EXCHANGE
APPLICATIONS

INTRODUCTION

As discussed in previous chapters, the development of ion exchange processes is closely related to the development of ion exchange products. This is due to the fact that each new development increased the applicability of the ion exchange unit process by:

- Increasing the available capacity.
- Increasing the range of application.
- Increasing the stability and operating life.

In addition, ion exchange, as a unit process, found rapid acceptance in the chemical engineering field, because in most cases it was found to be:

- More cost effective than wet chemical processes in a given application.
- Capable of producing a purer recovered product.
- Capable of producing high-purity water, at a fraction of the cost of distillation, in the large volumes required by industry.

Ion exchange, however, is not the "magic wand" that solves all water-treatment problems. Each new application must be evaluated on its own merit. With this in mind, it is the counter-ion that is exchanged for other ions in the water or solution being processed. The fixed polar group does not, in reality, take part in the exchange reaction. The fixed polar group does, however, influence the type of ion exchange reaction that can occur in a given application, and affects the relative selectivity for the several ionic species which may be present in the media being treated.

Some familiar and not so familiar applications of ion exchange will be discussed in the sections which follow. The applications discussed are based upon the properties of current commercially available ion exchange products.

WATER TREATMENT

The key to the economical use of the ion exchange process is the fact that ion exchangers can be regenerated, and that in most applications they will operate without problems for many thousands of cycles. Regeneration is accomplished by passing a small volume of the regenerant chemical at high concentration downflow through the bed contained in a suitably designed unit. This by its very nature is a batch process. The regenerant concentration is always much higher than the process stream. This technique has been in use since the first industrial application, and is now known as "co-flow regeneration." At exhaustion (end of service) the ion exchanger will contain relatively high quantities of those ions removed during the service run. The regeneration process removes these ions from the partially exhausted resin bed and restores the resin to its appropriate ionic form for the next cycle. In "co-flow" regeneration, small amounts of the process water ions remain in the lower portion of the ion exchanger bed. These small residuals will show up in the process stream during the next service cycle as "leakage." The amount of "leakage" is dependent on the regeneration efficiency and the equilibrium constants which apply in any given application. Usually, these low "leakage" levels do not limit the ion exchange process. However, regeneration techniques needed to be modified for those applications where "leakage" in the process stream could not be tolerated. The equilibria at the effluent end of the ion exchange bed needed to be changed, and several methods were tried:

- decreasing the flow rate to allow more contact time;
- increasing the regenerant concentration to drive the equilibria in the desired direction;
- increasing the regenerant quantity (dosage);
- Air-mixing the regenerated bed to distribute the residuals and to reduce their impact on service cycle;
- Regenerating upflow to ensure that the lower portion of the bed would be fully regenerated.

Upflow regeneration was found to be a very cost-effective answer to the "leakage" problem when the movement of the ion exchanger bed was restricted during the upflow regeneration. This regeneration technique is known as "counterflow regeneration." With this innovation "leakage" levels can be markedly reduced and better effluent qualities are achieved without increased operating costs.

CATION EXCHANGE APPLICATIONS

Sodium Cycle—Softening

Natural zeolites, synthetic aluminosilicates, and currently available high capacity polymer products are known to remove Calcium and Magnesium ions from natural water supplies when regenerated with ordinary salt. This process is commonly known as "softening" and the reactions involved are illustrated by equations 5-1 (exhaustion) and 5-2 (regeneration). In these equations, R presents the ion exchanger matrix. Note that the ion exchange reactions are reversible and that the regeneration allows one to repeat the softening operation over many cycles. This is a basic property of all ion exchange applications.

Service Cycle

$$
\begin{matrix} Ca(HCO_3)_2 \\ MgSO_4 \\ NaCl \end{matrix} + 2R\text{-}Na \rightleftharpoons R_2 \left\langle \begin{matrix} Ca \\ \\ Mg \end{matrix} \right. \quad \begin{matrix} 2NaHCO \\ + \; Na_2SO_4 \\ 2NaCl \end{matrix} \tag{6-1}
$$

Regeneration Cycle

$$
R_2 \left\langle \begin{matrix} Ca \\ \\ Mg \end{matrix} \right. + 2NaCl \rightleftharpoons 2R\text{-}Na + \begin{matrix} CaCl_2 \\ \\ MgCl_2 \end{matrix} \tag{6-2}
$$

Since only neutral or slightly alkaline water supplies could be softened by the natural zeolites and synthetic sodium aluminosilicate products, this process was once called "base exchange"; however, that term is seldom used today.

Hydrogen Cycle—Dealkalization

The ion exchange product of choice for the removal of alkalinity from water supplies is the weak acid carboxylic $(R-COOH)$ resin. The first ion exchange material used industrially in this process was sulfonated coals. Currently, polymeric products are used, and these materials have a much higher capacity and, therefore, are more cost-effective. When water supplies are used without treatment in boilers, "boiler scales" form which are difficult to remove, boiler tubes fail, and heat transfer is markedly reduced. Dealkalization of the water supply before use almost eliminates most of the scaling problem since the process reduces both the alkalinity and the total hardness, which produces the adherent scales on the boiler tubes and other heat-transfer surfaces. The weak acid cation process of "dealkalization" is illustrated by equations (6-3) and (6-4).

$$
Ca(HCO_3)_2 + 2(R-COOH) \rightleftharpoons (R-COO)_2 \left\langle \begin{matrix} Ca \\ \\ Mg \end{matrix} \right. + \begin{matrix} NaHCO_3 \\ | \\ NaCl \end{matrix} \tag{6-3}
$$

$$
R-COOH + NaHCO_3 \rightleftharpoons R-COONa + \begin{matrix} Na_2SO_4 \\ | \\ H_2CO_3 \end{matrix} \tag{6-4}
$$

Note that:

- The bicarbonate alkalinity (HCO_3) is removed by converting it to carbonic acid (H_2CO_3), which can be reduced by degasification.
- The cations equivalent to the bicarbonate alkalinity are also removed so that a partial demineralization is accomplished.
- Weak acid cation exchangers are very selective for calcium and magnesium ions; therefore, run endpoints are determined by the appearance of sodium bicarbonate in the effluent stream.

Regeneration is accomplished by using dilute (1-2%) solutions of a strong acid, such as hydrochloric or sulfuric acid: this process is illustrated in equations (6-5) and (6-6), which use sulfuric acid.

$$2(R-COO) \left\langle \begin{matrix} Ca \\ \\ Mg \end{matrix} \right. + H_2SO_4 \rightleftharpoons 2R-COOH + \left. \begin{matrix} Ca \\ \\ Mg \end{matrix} \right\rangle SO_4 \qquad (6\text{-}5)$$

$$2R-COONa + H_2SO_4 \rightleftharpoons 2R-COOH + Na_2SO_4 \qquad (6\text{-}6)$$

These reactions are equilibria; and in the absence of free hydrogen ions calcium, magnesium and sodium will be preferred by the exchange site. When there are hydrogen ions available, the reverse reaction will occur. In fact, when the pH is low enough (<5), there will be almost no calcium, magnesium, or sodium removal.

ANION EXCHANGE APPLICATIONS

Anion exchange products are not widely used for domestic water-treatment applications. The reason for this is that the anions commonly associated with domestic water supplies are not objectionable for ordinary household use, and the removal will not improve the usefulness of the water supply. Strong base anion exchangers have been used to remove nitrates from domestic water supplies, since nitrates constitute a health hazard. The reaction involved in this application is:

$$R-Cl + \begin{matrix} SO_4^{-2} \\ \\ NO_3^{-1} \end{matrix} Cl^{-1} \qquad R \left\langle \begin{matrix} SO_4 \\ \\ NO_3 \end{matrix} \right. + Cl^{-1} \qquad (6\text{-}7)$$

As this equation shows that both the sulfate and the nitrate ions are removed, the ion exchanger in fact prefers the sulfate ion because of its double valence. While this process is useful in certain areas, it is not widely applied.

Dealkalization can also be carried out using a strong base anion exchanger. This process is, in general, much less efficient, since it requires a higher regenerant dosage than the weak acid cation resins in the same application.

INDUSTRIAL APPLICATIONS

Deionization and Demineralization

Deionization of water supplies became possible when strong acid cation exchangers and weak base anion exchangers were made commercially available (see Chapter 2). When these two products were used in series, almost complete deionization became a possibility. The cation exchanger in the hydrogen form removes the cationic species and forms the corresponding free acids [equation (6-8)]. The anion exchanger in the free base form absorbs the free acids leaving the process water free of ionizing species [equation (6-9)].

$$\begin{matrix} NaCl & & R-SO_3Na & HCl & \\ Ca(HCO_3)_2 & + \ R-SO_3H \ \rightleftharpoons & (R-SO_3)_2Ca & + \ H_2CO_3 & (6\text{-}8) \\ MgSO_4 & & (R-SO_3)_2Mg & H_2SO_4 & \\ Na_2SiO_3 & & & H_2SiO_3 & \end{matrix}$$

$$\begin{matrix} HCl & & R-N(CH_3)_2-HCl & \\ H_2SO_3 & + \ R-N(CH) \ \rightleftharpoons & R-N(CH_3)_2-H_2CO_3 \ + \ HCO_3^- & (6\text{-}9) \\ H_2SO_4 & & R-N(CH_3)_2-H_2SO_4 & \\ H_2SiO_3 & & H_2SiO_3 & \end{matrix}$$

The above equations clearly indicate that all ionic species in the water supply can be removed by ion exchange processes. Weakly ionized species (i.e., H_2CO_3) are partially removed, while very weakly ionized species (i.e., H_2SiO_3) are essentially unaffected by the ion exchange process. That is why the process has been named "deionization" as contrasted to "demineralization." In this two-bed deionization process, it is usual to regenerate the cation exchange material with a strong acid, and the weak base anion exchange material with ammonium hydroxide or sodium carbonate. As a process, "deionization" was applied industrially to boiler water treatment. It was extended to the de-ashing of crude sugar solutions in order to improve crystalization yields and lower the cost of sugar production.

   It is understood from previous chapters that the basic strength of an anion exchange product depends on the type of amine attached to the polymer matrix. Anion exchange products, which have quaternary amine groups attached, are completely dissociated and behave as strong bases, similar to strong acid cation exchange materials. Those ion exchange products, which have type-I or type-II quaternary groups, are in reality the only materials that can be described as true anion exchangers. Weakly basic anion exchangers are really acid absorption media and do not actually exchange ions.

Deionization is accomplished when a strong acid cation exchanger in the hydrogen form is used in series with a strong base anion exchanger in the hydroxyl form to treat a water supply. This process is illustrated by equations (6-10) and (6-11), as follows:

$$NaCl \qquad\qquad\qquad\qquad R-SO_3Na \quad HCl$$
$$Ca(HCO_3) + R-SO_3H \rightleftharpoons (R-SO_3)_2Ca + H_2CO_3 \qquad (6\text{-}10)$$
$$MgSO_4 \qquad\qquad\qquad (R-SO_3)_2Mg \quad H_2SO_4$$
$$Na_2SiO_3 \qquad\qquad\qquad R-SO_3Na \quad H_2SiO_3$$

$$HCl \qquad\qquad\qquad\qquad R-N(CH_3)_3Cl$$
$$H_2CO_3 + R-N(CH_3)_3OH \rightleftharpoons R-N(CH_3)_3CO_3 \qquad (6\text{-}11)$$
$$H_2SO_4 \qquad\qquad\qquad R-N(CH_3)_3SO_4$$
$$H_2SiO_3 \qquad\qquad\qquad R-N(CH_3)_3SiO_2 + H_2O$$

These equations represent ideal reactions, but in the real world small inefficiencies (leakages) will yield a treated water with some residual ions. This effect can be reduced to some extent by:

- Using uneconomical high regenerant dosages.
- Reducing the flowrate to allow more contact time for the reactions.
- Using unflow regeneration to ensure that the bottom of the ion exchanger bed is fully regenerated.

In 1951, R. Kunin and F. X. McGarvey introduced the concept of using the regenerated strong acid cation exchanger and the strong base anion exchanger in a single mixed bed. While a reduction in volume throughput did occur, the technique markedly improved the water quality that could be achieved for most water supplies. The reaction which takes place in a mixed bed when treating a solution of a neutral salt (i.e., sodium chloride) is shown in equation (6-12).

$$R-N(CH_3)_3OH$$
$$+$$
$$R-SO_3H + NaCl \rightarrow R-SO_3Na + HCl \qquad (6\text{-}12)$$
$$\downarrow$$
$$R-N(CH_3)_3Cl$$
$$+$$
$$H_2O$$

Similar reactions can be written for other salts commonly found in water supplies.

While leakage cannot be entirely eliminated, inspection of equation (6-11) reveals that the overall reaction in a mixed bed is driven to completion because one of the endproducts is undissociated water. As a result, the quality of the treated water from a mixed bed is generally better than that produced by a two-bed system using the same ion exchange materials. The mixed-bed process of demineralization has been producing treated water equal to double distilled water for a wide range of applications since 1951, at a fraction of the cost of distillation. Some typical treatment systems which have been used industrially to produce pure, and ultrapure water, in quantity are illustrated in Figure 6-1.

Regeneration of a mixed bed is somewhat more complicated than a single bed. The anion and cation exchange materials are intimately mixed and must be separated before the regenerant solutions are passed through the bed. The separation is usually accomplished by backwashing the exhausted mixed bed. As shown in Table 6-1, the relative densities and settling velocities of the exhausted anion and cation exchangers are different enough to provide good separations in most cases. After separation, the anion and cation exchange materials can be regenerated separately or simultaneously.

The separations achieved by backwashing alone are, however, not perfect. Since ion exchange materials have particle size distributions which range from 16 mesh (1.19 mm) to 50 mesh (0.297 mm), absolute separations are not easily accomplished. After backwashing, some fine particles of the cation exchanger will be entrained in the lower portion of the separated anion section, and some coarse anion exchanger will be entrained in the top portion of the separated cation section. When the regeneration is completed, the entrained anion exchange material will be in the SO_4 form; while the entrained cation exchange material will be in the Na form. During the service cycle, these ions are available for exchange and, as a result, the treated water quality will not be as high as theoretically possible.

Several commercial products and process modifications are now available to improve the performance of mixed beds for water treatment. Inert bead polymers with densities > 1.09 and < 1.30 are being used to form a barrier between the separated anion and cation exchange materials during the backwashing process. While these products reduce the degree of cross-contamination, they do not eliminate the effects entirely.

For those applications in which low level ion leakage is an important consideration, external regeneration of the mixed bed is now being employed. These systems use water to sluice the exhausted mixed bed to especially designed equipment where separation, regeneration, and remixing can be carefully controlled.

- Separation is achieved in one unit where the mixed bed is backwashed with a solution of sodium hydroxide. The density of the sodium hydroxide is high enough to float the anion exchanger to another unit to

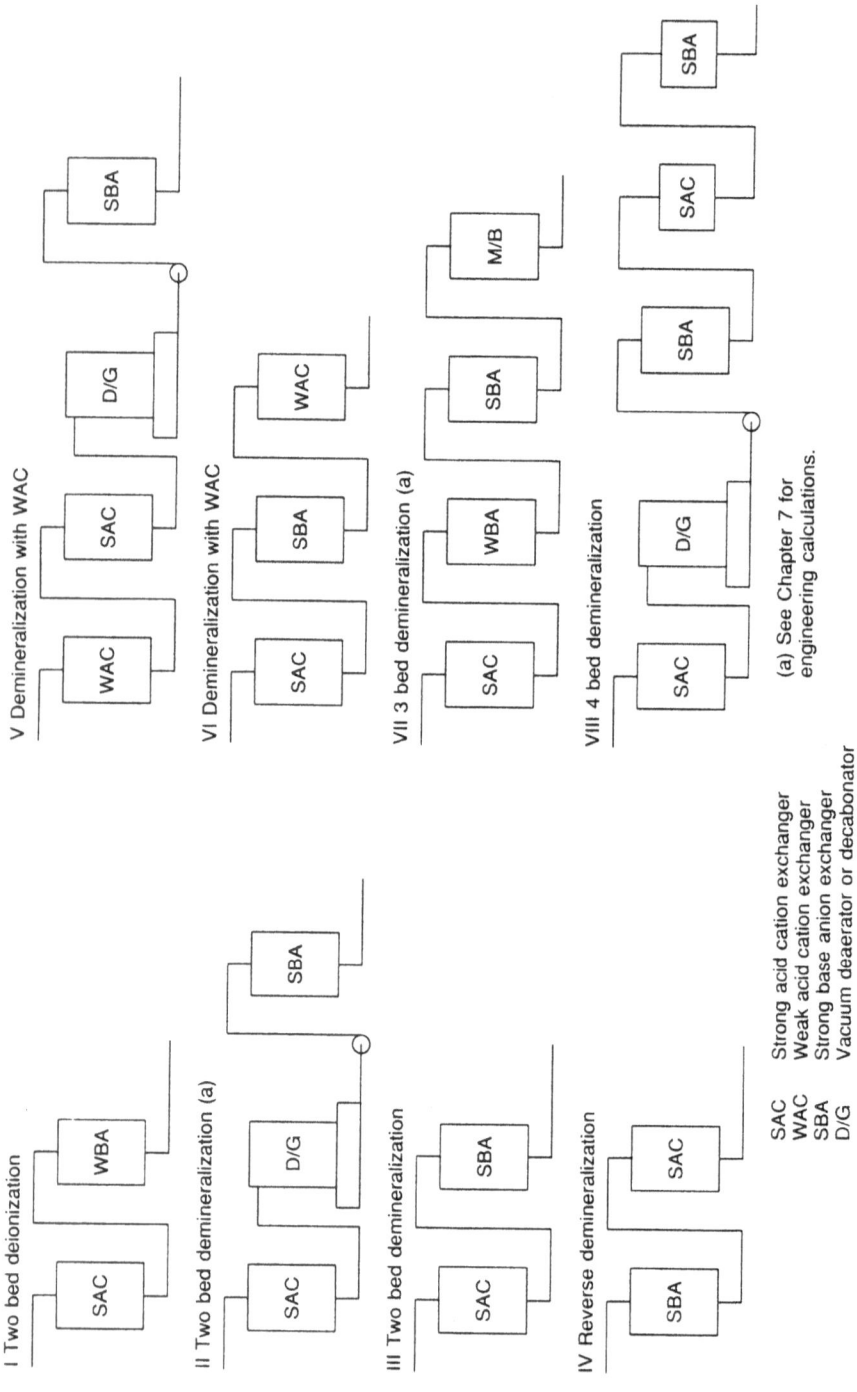

Figure 6-1. Deionization and demineralization.

I Two bed deionization

II Two bed demineralization (a)

III Two bed demineralization

IV Reverse demineralization

V Demineralization with WAC

VI Demineralization with WAC

VII 3 bed demineralization (a)

VIII 4 bed demineralization

(a) See Chapter 7 for engineering calculations.

SAC Strong acid cation exchanger
WAC Weak acid cation exchanger
SBA Strong base anion exchanger
D/G Vacuum deaerator or decabonator

IX Two bed system with M/B douche

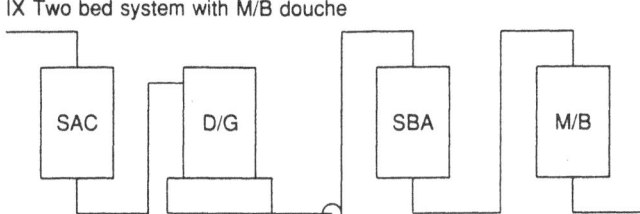

X Mixed bed demineralization

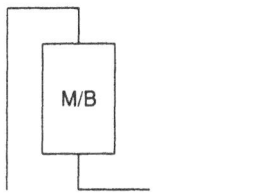

XI Mixed bed with recycle

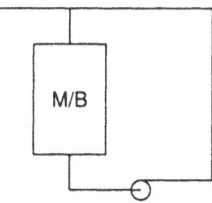

XII Stratified demineralization system

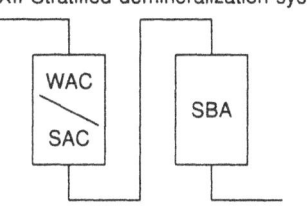

Figure 6-1. (*Continued*)

Table 6-1. Separability of Ion Exchange Materials

Mixed Bed Component	Anion Exchanger	Cation Exchanger
Exhausted ionic form	Cl	Na
Density, g/cc	1.09	1.30
Settling velocity, cm/sec[a]	3.98	12.95

[a]Calculated using Stoke's law and a mean particle diameter of 0.42 mm (~40 mesh).

complete the anion regeneration and final rinse. During this process, the cation exchange material which remains in the first unit, is almost entirely converted to the sodium form.

- The Na form cation exchange material, now free of anion resin particles, is regenerated with a strong acid and rinsed.
- The regenerated and rinsed components are sluiced, using treated water, to a third unit where they are air-mixed and stored until needed.

This type of remote regeneration process ensures that the mixed bed is fully regenerated, and keeps the Na and SO_4 leakages to a very low level during the service cycle. Systems of this kind are being used for condensate polishing treatment in "once-through" steam generator plants, and in "boiling water reactor" plants. These applications will be discussed in later sections of this chapter.

The cost of operating a demineralizer system is directly related to the quantity of total dissolved solids in the water supply to be treated. Assuming that regenerations are carried out every 8 hr, the upper limit for the dissolved solids can be estimated as follows:

$$\frac{13 \text{ kg/ft}^3 \times 1000 \text{ g/kg} \times 17.1 \text{ mg/g}}{2 \text{ gal/min ft}^3 \times 8 \text{ hr} \times 60 \text{ min/hr} \times 3.785 \text{ l/gal}} = 80 \text{ mg/L}$$

Therefore, to control operating costs, some alternate scheme is needed when the water supply to be treated contains a few hundred mg/L of dissolved salts.

One technique being used in the water-treatment industry is a combination of reverse osmosis and demineralization. Reverse osmosis is a process that uses pressure and a membrane to exclude dissolved species from the product (permeate) stream, and concentrate those species in a reject stream. Typical performance of reverse osmosis are:

Total flow 100 gal/min
Reject flow 10 gal/min
Permeate flow 90 gal/min
Salt rejection 90% or more

The attractive feature of reverse osmosis is that often the reject stream containing 90% of the dissolved solids can be returned to the environment, since no chemicals have been added in the process. Starting with a water supply containing 500 mg/L of total dissolved solids, the following performance is possible, based on a total flow of 100 gal/min:

	First Stage	Second Stage
Reject stream, gal/min	10	9
mg/L	4500	550
Permeate stream, gal/min	90	81
mg/L	55.5	6.2

Thus, single-stage reverse osmosis will yield a process stream that can be demineralized economically; and two-stage reverse osmosis would reduce the regenerant costs by a factor of almost 9.

Reverse osmosis, therefore, is a useful unit operation in the water-treatment field, and commercially available from small "household" units to large installations, capable of handling thousands of gallons per minute.

CONTINUOUS ION EXCHANGE

Continuous ion exchange systems are characterized by the fact that the processes of backwash, regeneration, rinse, and service are carried out simultaneously. As a result, the equipment is designed so that each step is isolated. In addition, the ion exchange material is continuously moved from treatment section to treatment section, in a direction which is counter to the direction of fluid flow. Three advantages are achieved by using continuous systems:

- Less ion exchange material is required compared to a typical batch-column operation.
- Typically, very high regeneration efficiencies are achieved with reduced regenerant dosages.
- Usually, the floor space required for a continuous system is less than that needed for a column-operating system.

There are some disadvantages that must be considered when continuous ion exchange is applied industrially. A uniform flow of ion exchange material from section to section is required to ensure that the treatment exchange zone is essentially undisturbed. The ion exchange material must have sufficient mechanical strength to withstand the periodic or continuous movement required by this application. A more or less sharp exchange zone must be maintained with variations in feedwater composition; this requires special controls and careful maintenance.

A number of continuous ion exchange processes have been described in the literature. In some, the ion exchange material moves from top to bottom. In others, the ion exchange material is sluiced from unit to unit and operates in each stage as a fixed-bed column. Fluidized beds have also been used; they operate in a single column in which perforated trays are employed to separate the various treatment stages.

Interest in continuous ion exchange treatment systems remains high because of the inherent advantages; however, only a few have reached industrial application, and many others have only attained the pilot plant stage.

The design of continuous ion exchange treatment systems, in general, is based on theoretical considerations. Equilibria and kinetic relationships are determined experimentally for a given application. After evaluation, the height of a theoretical plate and the mass transfer coefficients are obtained. The results are then used to size the equipment and establish the flowrates for each process section. There are currently three systems which have achieved industrial status, and these will be discussed below.

THE HIGGINS PROCESS

The Higgins process, which dates back to the 1950s, is shown schematically in Figure 6-2. The unit has several functional sections isolated from each other by butterfly valves. In each treatment section, the ion exchange material moves counter-flow to the fluid being passed through that treatment section. Actually, the system is not truly continuous, since the ion exchanger is moved in discreet amounts at frequent intervals and the various process fluids do not run when the resin is moving. Since the ion exchange material is moved in small volumes

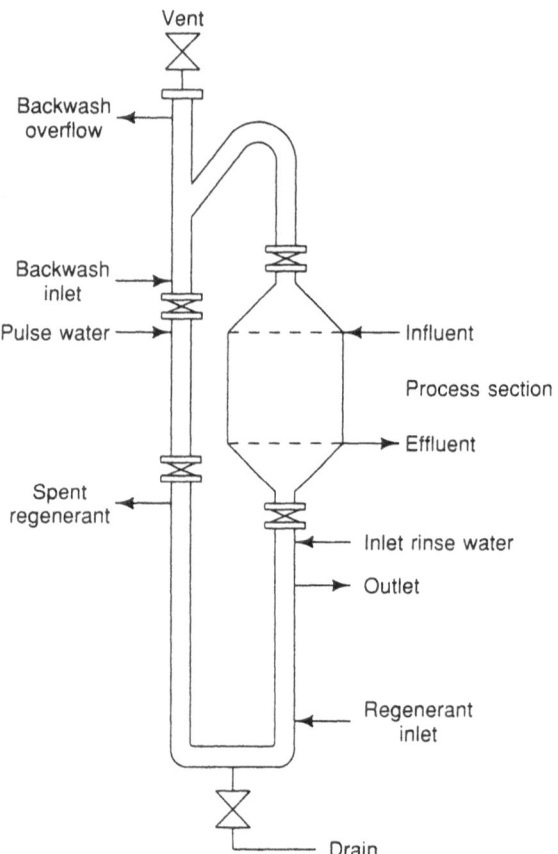

Figure 6-2. Higgins continuous ion exchange system.

and at frequent intervals, the overall effect is one of continuous treatment. Each process zone is optimized for the process taking place in it. As Figure 6-2 shows, the system consists of treatment zone, backwash zone, regeneration zone, and rinse zones—just like a single-column operation. In addition, there is a pulse section which uses water to move the ion exchanger from section to section. The advantages of the Higgins process are that the resin flows counter to the processing fluid in all sections. Therefore, the ion exchanger leaving the treatment section is close to complete exhaustion; and the ion exchanger entering the treatment section is almost completely regenerated and rinsed. The fluid being treated, therefore, comes in contact with fully regenerated ion exchange material as it leaves the unit. The same is true for the regeneration section; and the overall operation is more efficient than the conventional fixed-bed column operation. The floor space requirements of a Higgins system is only 20–25% of that of a fixed-bed installation but requires more height. The height has been accommodated in some installations by putting the system in a pit. The Higgins process does have a disadvantage, in that every time the butterfly valves are closed to isolate the treatment sections, a small amount of the ion exchange material is crushed. This can become critical after a relatively short operating period. Therefore, it is important that ion exchange materials used in this system have a high mechanical strength and an initial high whole bead count.

THE ASAHI PROCESS

The Asahi process was developed in Japan to overcome some of the disadvantages of the Higgins process equipment. In this process, the treatment stages have been separated into different vessels. In principle, the system consists of a treatment vessel, a regeneration vessel, and a wash vessel as shown in Figure 6-3.

The ion exchange material is moved from unit to unit by controlled pressure changes. This system uses plug valves to isolate the treatment sections, and as a result is somewhat less damaging to the ion exchanger. The Asahi process has been successfully applied to the softening of water supplies. In addition, deionization is possible when two systems are operating in series.

THE FLUICON PROCESS

The Fluicon process uses the fluidized bed principal to accomplish continuous ion exchange treatment. As shown in Figure 6-4, the flow of ion exchange material from section to section is accomplished by hydraulic differences; and there are no isolation valves between stages. As a result, this system imposes less stress on the ion exchanger than either the Higgins or the Asahi process. Since the ion exchanger is fluidized in each stage, the system is not as efficient

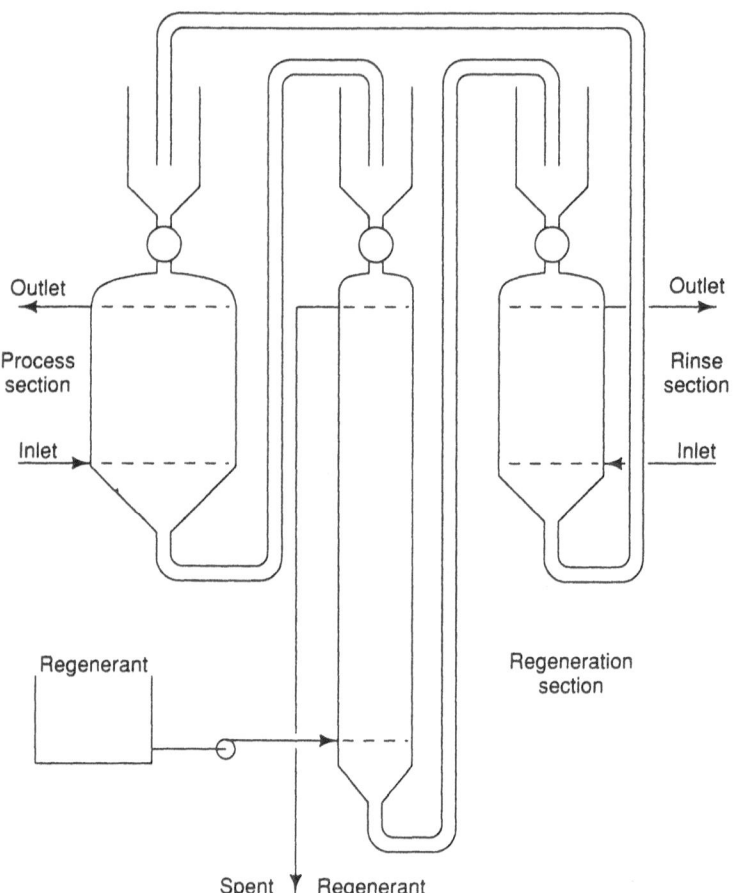

Figure 6-3. Ashai continuous ion exchange system.

as a compact-bed operation. The Fluicon process has been applied to the softening of water supplies in Germany with some success. The space requirements for this application is comparable to the Higgins system and smaller than the Asahi system.

The Higgins and Asahi systems have been the most successful in industrial water treatment. The Fluicon process, however, is useful in the field of hydrometallurgy for the treatment of ore slurries and waste streams. With the required automatic controls, these continuous ion exchange treatment systems represent considerable capital investment. Simpler single-unit fixed-bed columns, while less efficient, will often yield the same effluent quality at a lower capital cost.

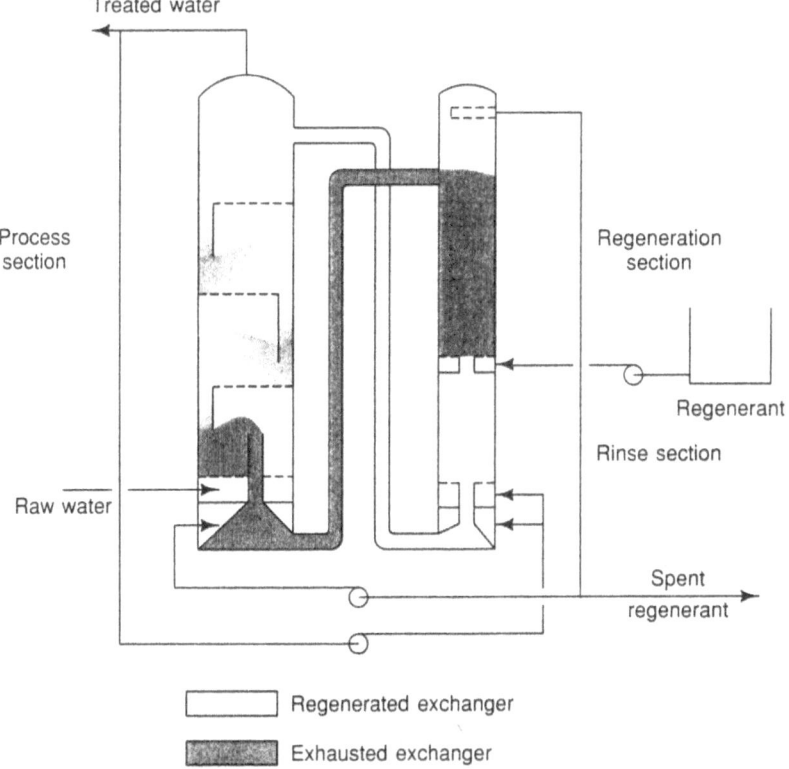

Figure 6-4. Fluicon continuous ion exchange system.

ELECTRODIALYSIS

The electrodialysis process uses an electric current to remove ions from solution. This separation is based on the use of two ion exchange membranes—one cation-permeable and one anion-permeable. The cation-permeable membrane allows the passage of cationic species such as Na^+, Mg^{2+}, or Ca^{2+} ions and low molecular weight cationic organics, while blocking the passage of anions and nonelectrolytes. The anion-permeable membrane allows the passage of anions such as Cl^-, SO_4^{2+}, or NO_3^-, while blocking the passage of cationic species and nonelectrolytes. In application, the cation and anion membranes are usually alternated in a plate and frame heat-exchanger-type configuration, commonly known as a stack. The direct current is introduced through two electrodes (anode and cathode), which drive the mobile ions in the appropriate direction. The compartments are connected to create a demineralized stream and a concentrated brine stream from a common feed. A simplified schematic of this

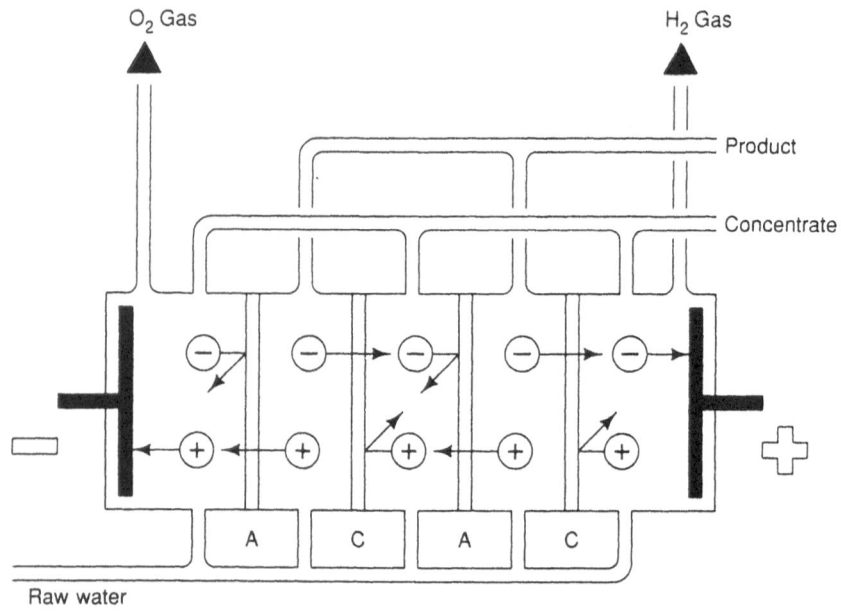

A = Anion permeable membrane
C = Cation permeable membrane

Figure 6-5. Typical ion exchange membrane stack.

process involving several compartments is shown in Figure 6-5. An electrodialysis system can be designed for almost any application involving ion exchange. The unique feature of this process is that it requires no regenerant chemicals and is, therefore, for the most part, less threatening to the environment than conventional ion exchange processes. The electrodialysis process is energy-efficient because the ion concentration is a small fraction of the total feed stream being processed.

Electrodialysis systems operate at low pressure (< 30 psi), with pumping only to circulate the process stream through the stack. This is in contrast to the reverse-osmosis process, which requires high pressure to separate the ions from the water phase. Some of the industrial applications which employ the electrodialysis process are:

- Concentration
    - —Seawater concentration for table salt.
    - —Recovery of electroplating waste.
    - —Recovery of useful salts from process effluents.
- Desalination
    - —Desalination of brackish water.
    - —Desalination of waste water.

—Demineralization of cheese whey.
—Purification of enzymes, vitamins, etc.
—Purification of sugars.
—Purification of amino acids.
—Purification of sera and vaccines.
—Treatment of electroplating wastes.
—Treatment of radioactive wastes.
- Chemical Reactions
    —Adjustment of acidity in fruit juices.
    —Production of organic acids from their salts.
- Dialysis
    —Treatment of aluminum etching waste liquor.
    —Acid separation in smelting processes.
    —Acid separation in metal manufacturing.
- Electrolysis
    —Electrolytic reduction and oxidation of organic compounds.

APPLICATIONS IN THE ELECTRIC POWER INDUSTRY

Throughout the electric power industry, water is the media of choice to transfer energy from the heat source; and in the form of steam, it drives the associated turbine/generators to produce useful electric energy. In most cases, the water supply is treated by ion exchange before use. Also, some power generation systems recycle and reuse a large fraction of the low pressure steam within the generating plant. A large portion of the worldwide electric power production comes from fossil fuel steam generation systems. A smaller portion is obtained from nuclear steam generation facilities. There are two types of nuclear steam generation systems in common use: the Pressurized Water Reactor (PWR) and the Boiling Water Reactor (BWR). The application of ion exchange materials in each of the above systems will be reviewed in the sections which follow.

Fossil Fuel Power Systems

The ion exchange applications reviewed here are also valid for gas-, coal-, or oil-fired boilers, and are typical of the supercritical steam generation systems in use today.

Since all the water entering a supercritical boiler is converted to steam, and after superheat is fed directly to the high pressure inlet of the turbine, the concentration of dissolved ions in this recirculating loop is of considerable importance. Recirculation in this closed loop is accomplished by condensing the low pressure steam from the turbine and returning it to the boiler through make-up heaters with booster pumps. In order to ensure that this recycled water is within plant specifications, all the water is treated by specially designed, high flow (50 gal/min/ft^2) condensate polishers. These units usually contain about 100 ft of fully regenerated mixed-bed ion exchange material in the H/OH form. At the flowrates employed, the pressure drop through the ion exchange bed is an important factor. Therefore, suppliers of these products often offer materials with somewhat coarser particle size than the conventional products. In addition, well-designed external regeneration systems are required to ensure the performance of the resin in the condensate polishing units. As expected, the make-up water for these generator systems is critical. While the make-up water volume may be small compared to the continually recirculating volume in the closed loop, it is also critical and must conform to strict quality standards, since all the dissolved species introduced will remain in the boiler tubes. Those species which may carry over (i.e., Na and SiO$_3$), with the steam can cause turbine operating problems—even at concentrations as low as 100 parts per billion or less. It is apparent, therefore, that the ion exchange processes play a very important role in the production of electric power in fossil fuel plants. A typical plant of this type, illustrating the ion exchange processes, is shown in Figure 6-6.

PWR Nuclear Steam Generator Systems

As the name (Pressurized Water Reactor) implies, the water which recirculates through the high temperature reactor core is maintained at high pressure to prevent boiling in the core. PWR plants are indirect generation systems; and the high pressure, high temperature primary coolant is passed through a "once-through steam generator" (OTSG), where heat is extracted by a secondary water system to produce steam which drives a turbine/generator, making useful electrical energy.

The concentration of dissolved and suspended solids in the recirculating primary coolant is kept as low as possible. In these systems, the dissolved species are generally introduced with the make-up water, and the quality of the make-up water is maintained by ion exchange treatment. The suspended solids, for the most part, result from the corrosion of the wetted internal surfaces. This corrosion is controlled somewhat by maintaining the primary coolant at a pH of 9.5–10.5, by using a side-stream purification process which uses mixed ion exchange beds in the NH$_4$/OH or Li/OH form. This side-stream purification is important since all dissolved and suspended species undergo neutron activation during recirculation through the core. As a result, any primary coolant leakage must be treated as a radioactive waste, requiring costly control and disposal.

The secondary water system extracts heat from the primary coolant in a OTSG, making steam for the turbine/generator sets. The low pressure turbine exhaust steam is condensed, purified by large condensate polishers, preheated, and returned to the inlet side of the OTSG. Thus, the secondary system is a closed loop, similar to the primary loop.

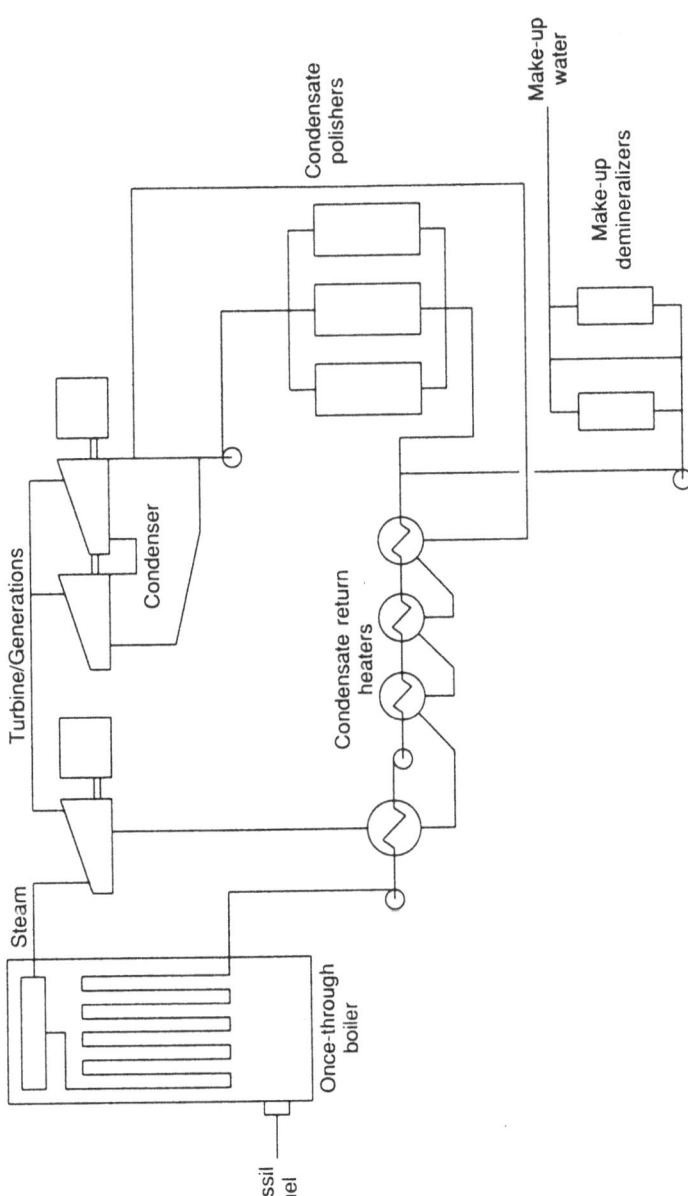

Figure 6-6.   Ion exchange applications in a fossil fuel plant.

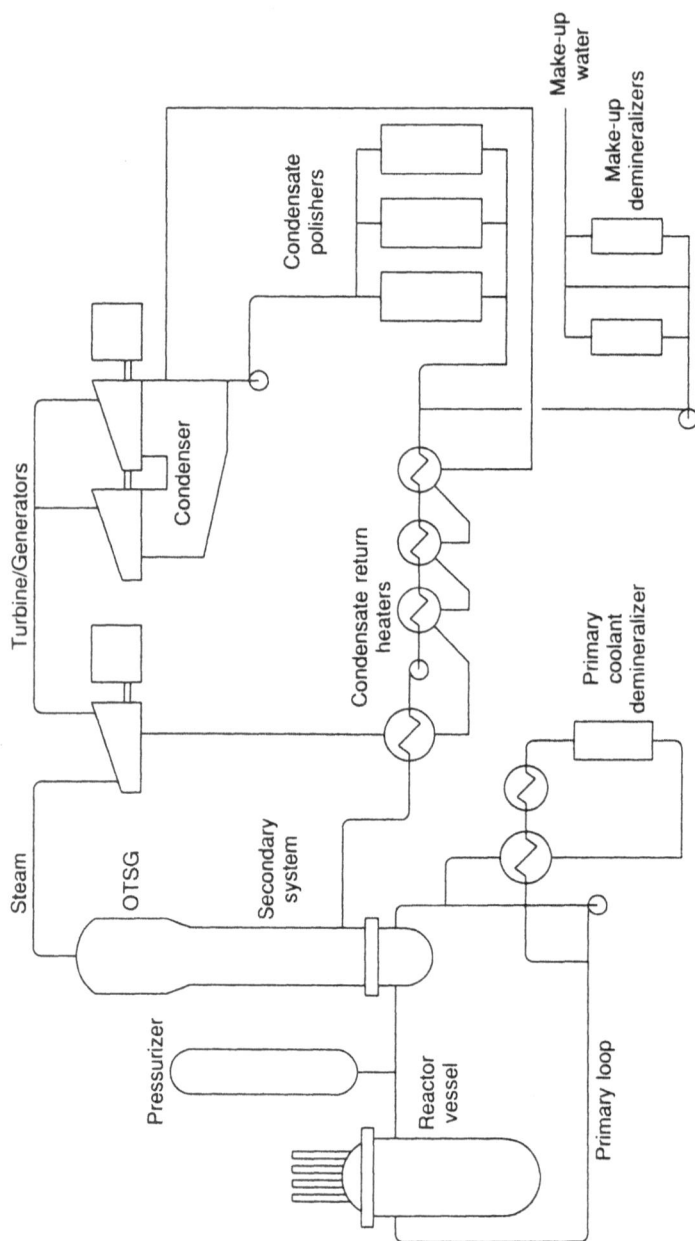

Figure 6-7. Ion exchange applications in a pressurized water reactor plant.

OTSG = Once Through Steam Generator

Chemical additives are often used to control corrosion in the preheaters, and reduce deposition in the OTSG, because these effects will reduce the overall plant efficiencies. Again, ion exchange is an important part of effective operation of a PWR plant which is being used for the production of useful electric energy. A typical PWR system, illustrating the ion exchange processes involved, is shown schematically in Figure 6-7.

BWR Nuclear Steam Generator Systems

In contrast to a PWR, a boiling water reactor (BWR) plant is a direct cycle system in which steam is produced in the high temperature core and goes directly to the turbine/generators, where it is converted to useful electric power. This type of system is a combination of elements from the fossil fuel and PWR plants. For example, the concentration of suspended and dissolved solids in the core water phase is controlled by using an ion exchange treatment system. In addition, the condensate is treated by a high flowrate polisher before being returned to the reactor core to produce additional steam. As previously illustrated, the ion exchange process is an integral part of a BWR power plant. A typical BWR system, showing the ion exchange processes involved, is given in Figure 6-8.

RECOVERY SEPARATION AND POLLUTION CONTROL

With the exception of the noble gasses (Krypton, Xenon, etc.), the ion exchange process has been applied to almost every element in the periodic table, and many of the synthetic nuclides beyond. These applications range from laboratory analytical techniques to full-scale production processes, and everything in between.

In the area of analytical chemistry, ion exchange is very useful since it will remove species from solution when the concentration of the species in the water is below analytical detectability. As a result, ion exchange is often used to concentrate and detect trace concentrations of toxic materials in rain water, surface and ground water supplies.

In the industrial area, we find ion exchange being used extensively for the recovery and concentration of valuable metals from dilute solutions. Also, the ion exchange process is being applied to the recovery of toxic metals from plant waste streams before they are discharged to the environment. There are several reasons for this:

- Ion exchange is a proven cost-effective process for recovering and concentrating species from dilute aqueous media.
- Regulation now requires manufacturers to remove toxic species from discharged water or face very stiff penalties.
- Manufacturers have found that the recovery, by ion exchange, of these otherwise wasteful metals is an economical process.

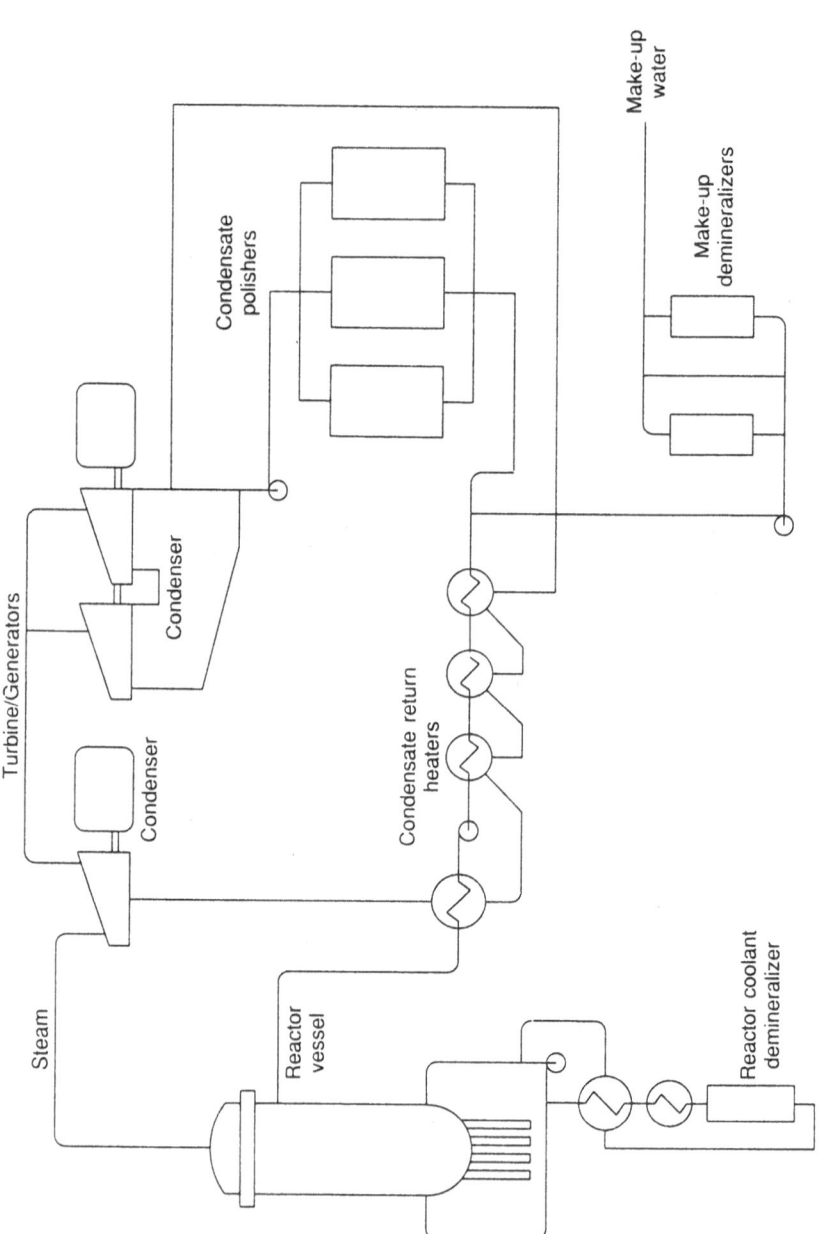

Figure 6-8. Ion exchange applications in a boiling water reactor plant.

A few examples of the removal of toxic metals from water supplies and industrial wastes are shown in Table 6-2. A careful literature search would turn up many more examples of this type.

Table 6-2. Selective Ion Exchange of Some Metals

Metal	Ionic Form	Resin Type[a]	Applicable Form	pH Range
Antimony	Antimoniate	SB	Cl	1–14
		WB	FB	4–10
Arsenic	Arsenate/arsenite	SB	Cl	1–14
		WB	FB	4–10
Barium	Divalent	WA	H or Na	1–8
Cadmium	Trivalent	WA	H	3–8
	Anion complex	SB	Cl	6–12
		WB	FB	6–12
Cesium	Monovalent	WA	Na	7–14
Chromium	Trivalent	SA	H	0–5
	Chromate anion	SB	SO_4	0–14
		WB	Cl	0–14
Cobalt	Divalent	SA	NH_4	1–8
		WA	H or Na	5–9
		WB	FB or SO_4	4–10
Copper	Divalent	WA	H or Na	5–8
		WB	FB or SO_4	3–10
Gold	Anion complex	SB	OH	0–14
Iron	Organic chelate	WB	FB	3–8
Lead	Divalent	WB	FB or SO_4	4–6
	Chloride complex	SB	Cl	0–14
Mercury	Chloride complex	SB	Cl	0–14
Molybdenum	Molybdate anion	WB	FB or SO_4	3–9
Nickel	Divalent	WA	Na or NH_4	5–9
		WB	FB or SO_4	4–7
Platinum	Anion complex	SB	Cl	0–14
Rhenium	Anion complex	SB	Cl	0–14
		WB	FB or Cl	0–14
Selenium	Selenite	WB	FB	2–6
Silver	Monovalent	SA	H or Na	0–7
		WB	FB	5–9
	Anion complex	SB	Cl	0–14
		WB	FB	0–14
Tungsten	Tungstate	SB	Cl	0–14
		WB	FB	0–14
Vanadium	Vanadate	WB	FB or SO_4	2–8
Zinc	Divalent	WA	Na	5–10
		WB	FB or SO_4	4–7
		SA	H	0–7

[a]SA, strong acid cation exchanger; SB, strong base anion exchanger; WA, weak acid cation exchanger; WB, weak base anion exchanger; FB, free base form.

Chapter 7

TECHNICAL DESIGN CALCULATIONS[1]

INTRODUCTION

The purpose a of water-treatment calculation is to determine the treatment system that satisfies the user's requirements at the lowest possible capital and operating costs.

In almost all applications, several schemes may have to be evaluated to find the most cost-effective treatment system. The data required for these calculations will include the following, and are usually defined by the end user:

- Treated water quality
- Quality of available water supply
- Run length and flowrate

STEPWISE PROCEDURE

Effluent Purity Required

The purity of the final treated water required by the enduser has a great influence on the complexity, capital cost, and operating cost of the designed system. As a result, it is uneconomical to design for a water quality greater than necessary. The range of water qualities required is very broad and depends on the industrial application as shown in Tables 7-1 and 7-2.

As indicated in Tables 7-1 and 7-2, a treatment system designed for a paper and pulp mill will be less complex than that required for a semiconductor manufacturer for the same water supply. The tolerance of the product for the various impurities in the final treated water must be considered when the system design is done.

Analysis of Water Supply

A well water supply, with adequate capacity, is preferable to that of a surface water supply, since well water in general is clearer, contains less organic mat-

---

[1]From *Demineralization By Ion Exchange*, Chapter 9, Academic Press, 1968.

Table 7-1. Water Quality Tolerances of Selected Industrial Applications (all values are given in mg/L unless shown otherwise)

Application	TH as CaCO	Alk. as CaCO	Fe as Fe	Mn as Mn	Turb.	Total Solids	pH
Carbonated							
beverages	250	50	0.2	0.2	0–2	<850	—
Breweries	>100	70–100	0.1	0.1	5–10	750	7.0
Ceramics	—	—	0.1	0.1	—	Low	—
Distilleries	—	—	0.1	0.1	—	5–10	—
Food processing							
Baking	—	—	0.2	0.2	5–10	—	—
Canneries	25–50	25–50	0.2	0.2	10	—	—
Candy	—	—	0.2	0.2	—	100	7.0
Starch prod.	—	—	0.1	0.1	—	—	—
Beverage ice	70	50	0.2	0.2	50	170	—
Laundries	0	—	0.1	0.1	0.1	—	—
Paper and pulp							
Fine paper	50	75	0.1	0.05	5	200	—
Kraft							
Bleached	100	75	0.2	0.1	5	300	7.0
Unbleached	200	150	0.5	0.3	25	500	7.0
Rayon pulp	8	50	—	—	5	100	—
Textiles							
General	75	100	0.1	0.1	5	200	—
Cotton	15	100	0.05	0.05	1.5	200	—
Rayon	10	75	0.05	0.02	1.0	200	—
Wool	0	70	0.1	0.1	5	200	—
Plastics	—	—	0.02	0.02	—	200	—

ter, and has a more constant composition. Such water supplies can often be used without pretreatment.

When a surface supply must be used, the variations in composition over one period must be available. This information is used as a basis for determining the average length of the service run and has an impact on the system's operating cost. A single analysis of a surface supply is often misleading, since the composition will be effected by seasonal rainfall. In some situations it may be

Table 7-2. Quality Tolerances of Ultrapure Water in Selected Applications (all values given in mg/L unless specified otherwise)

Industry	Fe as Fe	Cu as Cu	Total Solids	Resistance Mohm/cm	pH
Semiconductors	0.005	0.005	Trace	>18	7.0
Television tubes	0.005	0.005	Trace	2–5	7.0
OTSG feed water[a]	0.005	0.005	0.05	10	9.3

[a]OTSG, once-through steam generator.

advisable to bring in a new well, or run a pipeline from a nearby lake or pond rather than use a local supply which is subject to large seasonal variations. The cost of a new well or pipeline may be offset by savings realized in capital cost, pretreatment cost, and operating cost.

Types of Pretreatment

Once a water supply has been selected, the type of pretreatment and the need for pretreatment must be determined. The following guidelines may be helpful in this respect:

- When the total hardness and alkalinity are high, lime/soda presoftening is recommended.
- Surface waters with low total dissolve solids are usually clarified by coagulation and filtration without presoftening.
- When the organic content is high, coagulation in the presence of free chlorine is used, followed by an activated carbon bed to remove excess chlorine. This is required to protect the strong base anion exchanger from organic fouling.
- For those well waters containing manganese and iron, flocculation is used as a pretreatment, followed by filtration to avoid fouling the cation exchange resins.

The pretreatment selected will alter the analysis of the water supply. Once the dosages have been set, the changes to the analysis are calculated to provide a "design analysis" upon which the demineralizer design is based.

The changes which occur from pretreatment chemical additions are given in Table 7-3.

Types of Systems and Resins

The required water quality and the "design analysis" after pretreatment are the two criteria which determine the basic demineralizer system and the types of

Table 7-3. Water Analysis Changes from the Addition of 1 mg/L of Pretreatment Chemicals

	Reductions, mg/l			Increases, mg/L			
Added Chemical	Alk.	CO	Alk.	CO	SO	Cl	TH
Aluminum sulfate	0.45	—	—	0.40	0.45	—	—
Ferrous sulfate	0.36	—	—	0.31	0.36	—	—
Ferric sulfate	0.75	—	—	0.66	0.75	—	—
Chlorine	—	—	—	1.3	—	1.4	—
Hydrated lime	—	1.11	1.26	—	—	—	1.26
Sulfuric acid	0.95	—	—	0.84	0.95	—	—

ion exchange resins needed. There are 10 basic systems in current use for water treatment, and an application summary of each follows. (See also Chapter 6, Figure 6-1 for schematics.)

System 1: Two-Bed with Weak Base Anion Resin. This system is used for industrial plants requiring a reduction of the electrolyte to 2 or 10 mg/L but requiring no reduction of silica; it is not applicable for boiler feed-water purposes.

System 2: Two-Bed with Strong Base Anion Resin. This system will reduce the electrolyte concentration to 2 or 3 mg/L and the silica to 0.02 or 0.1 mg/L as silica.

System 3: Three-Bed. This system will save caustic soda regenerant with water supplies with a high total mineral acidity (sulfates and chlorides). It also often allows the omission of the decarbonator. The caustic soda is passed counterflow through the strong base and weak base anion exchangers, when the effluent conductivity of the weak base ion exchanger rises. The strong base at that time will only be partially exhausted; therefore, the latter often has enough excess capacity to remove the carbonic acid economically, even if the decarbonator is omitted. It also avoids early silica breakthrough.

System 4: Four-Bed with Primary Weak Base and Secondary Strong Base Resins. This system will reduce the electrolyte concentration to 0.2 or 1.0 mg/L and the silica to 0.02 or 0.1 mg/L. The use of primary and secondary cation exchange units save acid regenerant. This system is suitable for treating water supplies with high percentages of sodium ion and low percentages of alkalinity, which may cause high sodium leakages. Weak base resins also save caustic soda with water supplies, which have a high percentage of total mineral acidity. The secondary units in this system act as polishers and can be smaller than the primary units.

System 5: Four-Bed with Primary and Secondary Strong Base Resins. This system saves as much acid as does system 4, but the strong base resins in both the primary and secondary stages will reduce the silica to 0.01 or 0.05 mg/L. The system also avoids organic fouling of the secondary units.

System 6: Two-Bed/Four-Bed. This system is used when the output of demineralized make-up water must be increased for short periods of time (when oily condensate must be discarded). Then the primary and secondary stages of the four-bed system are operated in parallel as a large two-bed system. The secondary units are made the same size as the primary, because during the regeneration of the primary unit, the secondary unit handles the full load. Some two-bed/four-bed plants have been designed so that the freshly regenerated primary pair is always switched to the secondary position by valves and pipe interconnections. While this arrangement is intended to improve effluent quality, organic fouling of the secondary units causes problems. Therefore, it is best to keep the primary units always in the primary position, in order to protect the secondary anion exchangers from organic fouling.

System 7: Mixed Bed Alone. This system is used in smaller plants to save investment costs, but the practice increases operating costs. The resin capacity in mixed beds is usually assumed to the 80% or 85% of the same resins in a two-bed treatment system; therefore, more acid and caustic are required, for regeneration.

System 8: Cation-Bed, Decarbonator, and Mixed-Bed. This system will reduce the amount of acid and caustic needed when compared to a mixed-bed operating alone on water supplies with high alkalinity. The cation-bed also filters the water and thus acts as a pretreatment until, protecting the mixed-beds.

System 9: Two-Bed with Weak Base and Mixed-Bed. This system will reduce the electrolyte concentration to 0.04 or 0.10 mg/L and the silica to 0.02 or 0.1 mg/L. It saves as much acid as does system 8, but uses more caustic soda than does system 8 when treating water supplies with a high total mineral acidity.

System 10: Two-Bed with Strong Base and Mixed-Bed. This system will reduce the electrolyte concentration as low as system 9, but will take the silica concentration down to 0.01 or 0.05 mg/L. Having the strong base resin in both the primary unit and the mixed-bed protects the latter from organic fouling. The mixed-bed, acting as a polisher, is regenerated infrequently.

Regenerant Levels and Capacities of Ion Exchangers

The effluent total dissolved solids (TDS), specified by the enduser, defines the cation leakage that can be tolerated for the design. This leakage is used to establish the acid regenerant dosage of the cation exchanger. In the same way, the allowable silica leakage is used to establish the caustic regenerant dosage of the anion exchanger. The relative effects of regenerant dosage on leakage levels are given in Tables 7-4 through 7-8 for generic ion exchangers regenerated downflow. Manufacturers of ion exchange products supply similar information (on request), which is specific to their products.

   After the regenerant levels have been estimated from the data in Tables 7-4 to 7-8, the corresponding operating capacities are used for the system design.

   The above capacities are expressed in kg/ft^3 as $CaCO_3$, conversion to

Table 7-4. Strong Acid Cation Exchange Capacity and Leakage with
Sulfuric Acid Regeneration

Regeneration, lb/ft³		2.5		5.0		7.5	
		Capac.	Leak.	Capac.	Leak.	Capac.	Leak.
Alk.%	Na%	$\frac{kg}{ft^3}$	%	$\frac{kg}{ft^3}$	%	$\frac{kg}{ft^3}$	%
0	0	8.3	0	11.4	0	13.1	0
	25	8.3	5	11.4	2	13.2	0
	50	8.3	10	11.5	4	14.1	2
	75	8.5	24	12.6	14	15.0	8
	100	10.3	62	16.9	42	20.0	29
50	0	8.6	0	11.9	0	13.5	0
	25	8.9	0	12.3	0	14.1	0
	50	9.2	5	13.1	2	15.0	2
	75	9.5	15	16.4	8	16.9	6
	100	11.5	39	19.2	25	22.7	16
100	0	9.0	0	12.0	0	14.0	0
	25	9.8	0	13.2	0	15.1	0
	50	10.8	0	14.8	0	16.9	0
	75	12.1	0	16.8	0	19.7	0
	100	13.5	2	21.4	1	25.5	0

eq/L of bed can be accomplished as follows:

$$\frac{kgr}{ft^3} \times \frac{17.1 \text{ g}}{kgr} \times \frac{eq \text{ wt } CaCO_3}{50 \text{ g}} \times \frac{ft^3}{28.3 \text{ L}} = \frac{eq}{L \text{ (Bed)}}$$

when necessary.

In order to satisfy current antipollution regulations, the combined spent regenerants from any designed system must be neutral before being discharged to the environment. Most industrial facilities will have a holding tank where these solutions can be mixed and the pH adjusted before discharge. It may be necessary in some plants to adjust the regenerant dosages in order to reduce the cost of this final neutralization process.

The final adjusted regenerant levels and volumes are used to calculate the anion and cation operating schedules. Such a schedule will include the following steps:

Backwash
Regenerant dilution
Regenerant introduction
Displacement (slow) rinse
Final (fast) rinse
Service run

Table 7-5.   Strong Acid Cation Exchange Capacity and Leakage with
Hydrochloric Acid Regeneration

Regeneration, lb/ft³		2.5		5.0		7.5	
		Capac.	Leak.	Capac.	Leak.	Capac.	Leak.
Alk. %	Na%	$\frac{kg}{ft^3}$	%	$\frac{kg}{ft^3}$	%	$\frac{kg}{ft^3}$	%
0	0	11.9	1.5	18.9	0.4	23.7	0.2
	25	11.7	3	18.5	2	23.5	1.6
	50	12.2	6.5	19.4	3.9	23.8	3
	75	13.3	11.5	20.8	5.1	25.4	5.1
	100	14.9	29	22.9	13.1	27.9	7.9
50	0	12.4	1.0	19.5	0.5	25.1	0.4
	25	12.2	2.5	19.1	1.5	24.4	0.5
	50	12.6	4	20.2	1.5	24.8	1
	75	13.8	8	21.7	3.1	26.9	1.6
	100	15.3	20	23.9	5.9	28.9	2.2
100	0	12.7	0	20.4	0	25.7	0
	25	12.5	0	20.2	0	25.2	0
	50	13.1	0	20.8	0	25.9	0
	75	14.2	0	22.5	0	27.5	0
	100	16.0	2	24.7	1	29.7	0

Table 7-6.   Weak Acid Cation Exchange Capacity for Alkalinity Removal

$\frac{Hardness}{Alkalinity}$ Ratio	TMA, mg/L as CaCO[a]	Capacity, kg/ft³ at	
		75 F	55 F
0	5	12	8
	65	13	10
	130	16	10
	280	18	14
0.5	5	12	9
	65	15	11
	130	20	13
	280	22	17
0.9	5	13	10
	65	17	12
	130	23	16
	280	26	19
>1.0	5–280	50	40

[a]TMA, total mineral acidity.

Table 7-7.   Strong Base (Type-I) Anion Exchange Capacity

Dosage $\left(\frac{lb}{ft^3}\right)$	% Weak Acids[a]	Capacity $\left(\frac{kg}{ft^3}\right)$	Silica Leakage, mg/L				
			Influent % Silica[b]				
			10	30	50	70	90
4	25	10.8	0.05	0.19	0.30	0.41	0.53
	50	11.8					
	75	12.6					
6	25	12.2	0.02	0.05	0.08	0.12	0.14
	50	13.4					
	75	14.2					
8	25	13.4	0.01	0.03	0.05	0.07	0.09
	50	14.2					
	75	15.0					

[a] % Weak acids $= \dfrac{(SiO + CO) \times 100}{TEA}$

[b] % Silica $= \dfrac{SiO_2 \times 100}{TEA}$

TEA, total exchangeable anions.

The volumes and flowrates of the above steps is used to determine piping size, valve sizes, and pumping requirements, as well as the total amount of water needed for each stage of the treatment train.

Length of Service Run and Flowrate

The enduser flowrate requirement is used to determine the total ion exchange bed area, and the bed volume is dependent on the length of the service run between regenerations. For most applications, flowrates of 6-8 gal/min/ft^2 are acceptable. In polishing applications, units are often operated at 15-25 gal/min/ft^2, and condensate mixed-bed polishers are designed to operate at 50 gal/min/ft^2 and above.

At some facilities the service run length may be exactly 24 hr between regenerations, in order to reduce the labor cost for manually operated equip-

Table 7-8.   Weak Base Anion Exchange Capacity for Acid Removal

Chloride Ion as % of TMA[a]	Capacity (kg/ft^3)
0–40	20
40–60	19
60–80	18
>80	17

ment. When fully automated designs became available, regeneration every 4–8 h became possible. The increased frequency reduces capital cost and also reduces the volume of resin required to treat a given water supply.

In condensate mixed-bed polisher applications, the ion exchanger is never regenerated in place. When regeneration is required, the resin is transferred to auxiliary vessels for separation regeneration, rinsing, and remixing.

In applications involving the removal of radioactivity, ion exchangers are usually never regenerated; the spent resins are dewatered and solidified for burial.

Treated Water Storage and Regenerant Storage

Without treated water storage, the pretreatment system and the primary ion exchange units must be designed to provide for backwash, regeneration, and rinse requirements for all additional units in the treatment train. This may require a larger front-end system and will increase capital costs. This situation can be eased by use of a treated water storage tank to provide for the additional volumes and flows.

In addition, adequate regenerant storage will reduce operating costs, since it permits the purchase of regenerant chemicals in bulk rather than in small quantities. In large demineralization facilities, the savings in operating costs can be appreciable and should be considered in overall plant design.

Final Calculations

When all the above items have been satisfied, the final design calculations can be started—this done in reverse order, working from the outlet (treated water end) to the inlet (raw water supply). For example, in a two-bed system, the anion exchange unit would be designed first, followed by the cation exchange unit, and finally the pretreatment facility. This approach is necessary since the cation unit supplies water for backwashing, regeneration, and rinsing the anion unit. In addition, the pretreatment system supplies water for similar operations of the cation unit.

Sample Calculations

Four examples of design calculations are given here in some detail to illustrate the techniques required.

Case A: Chlorination and Lime Softening Pretreatment Followed by Two-Bed Demineralization with Deaeration (Figure 7-1).

1 Design Data. The total flow is 266 gal/min without demineralized-water storage for make-up to high pressure (2400 psi) boilers. The water supply is

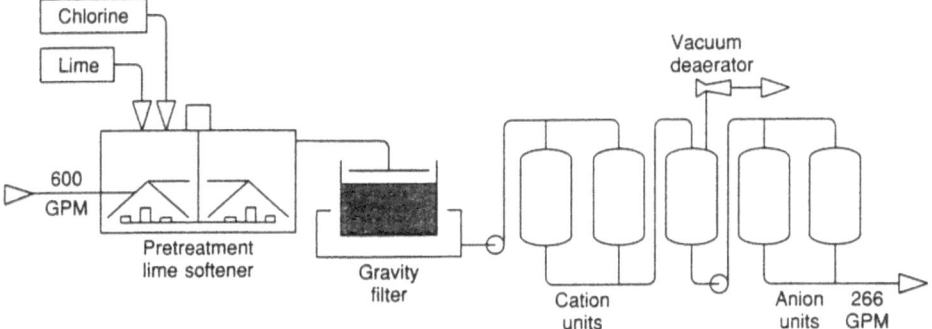

Figure 7-1. Case A: Final process schematic.

from a reservoir of fairly constant analysis as shown in Table 7-9. The final treated water quality is:

Total dissolved solids < 2 mg/L
Conductivity < 10 micromho
Residual silica < 0.03 mg/L as SiO

2 Modification of Water Analysis by Pretreatment. (a) The first step is to calculate the effect of the chlorine addition (5 mg/L is assumed for this case).

Table 7-9.   Case A: Estimated Operating Performance[a,b]

	A	B	C	D	E	F	G
Calcium	56	56	56	35	0	0	0
Magnesium	24	24	24	22	0	0	0
Sodium	55	55	55	55	2	2	2
Hydrogen	0	0	0	0	41	41	0
Total cations	135	135	135	112	43	43	2
Bicarbonate	108	101	92	0	0	0	0
Carbonate	0	0	0	69	0	0	0
Hydroxyl	0	0	0	0	0	0	1
Sulfate	12	12	21	21	21	21	0
Chloride	15	22	22	22	22	22	1
Total anions	135	135	135	112	43	43	2
CO_2 as CO_2 8	14.5	22.5	0	30	5	0	
Silica as SiO_2	7.4	7.4	7.4	7.4	7.4	7.4	0.03
pH	7.5	7.2	6.9	10.2	2.9	3.0	9.0
Turbidity JTU[c]	100			0	0	0	0
Color	30			5	5	5	5

[a]A. raw water supply: B. after chlorine addition; C. after ferric sulfate addition; D. after lime addition; E. effluent from cation exchanger: F. after deaeration: G. final treated water (anion exchanger effluent).
[b]Values are in mg/L as $CaCO_3$ unless shown otherwise.
[c]JTU. Jackson turbidity units.

As shown in Table 7-3, 1 mg/L chlorine addition increases the chloride ion by 1.4 mg/L, decreases the alkalinity by 1.4 mg/L, and increases the carbon dioxide by 1.3 mg/L. The changes to the water analysis is calculated by multiplying each of the above by the chlorine dosage used for the case. These changes are reflected in the design analysis shown in Table 7-9.

(b) The second step is to modify the analysis for the coagulation process. A dosage of 12 mg/L of ferric sulfate is selected for this lime-softening pretreatment. The ferric hydroxide floc will precipitate well at the high pH values present. Again, as Table 7-3 shows, 1 mg/L ferric sulfate decreases the alkalinity by 0.75 mg/L, increases the carbon dioxide by 0.66 mg/L, and increases the sulfates by 0.75 mg/L. After multiplication by the ferric sulfate dosage chosen, the resultant changes are reflected in the design analysis given in Table 7-9.

(c) The last step is to calculate the effect of the lime addition required. Since the bicarbonate alkalinity exceeds the calcium hardness, the lime dosage required is calculated as follows:

$$\text{Alkalinity} = 92 \text{ mg/L as CaCO}$$

$$CO_2 (22.5 \times 2.3) = 52 \text{ mg/L as CaCO}$$

$$\text{Magnesium} = \frac{24 \text{ mg/L as CaCO}}{168 \text{ mg/L as CaCO}}$$

$$\text{Lime} = (168 \times 74)/93 = 132 \text{ mg/L of 93\% Ca(OH)}$$

$$168/150 = 1.1 \text{ lb of 93\% Ca(OHO}_2/1000 \text{ gal)}$$

3 Design Calculations. (a) Definition of Demineralizer System. To produce a treated water quality with less than 2 mg/L dissolved solids will require a two-bed system with a strong base anion resin to assure that the residual silica of less than 0.03 mg/L can be satisfied. Since the alkalinity in the design analysis is 69 mg/L, a vacuum deaerator will be used between the cation exchange unit and the anion exchange unit. This unit will also reduce the dissolved oxygen to less than 0.5 mg/L. The estimated water analysis after each treatment stage is shown in Table 7-9.

(b) Selection of Resins, Regeneration Dosages, and Capacities. Now the selection of the ion exchange resins, their regenerant dosages, and the relative capacities can be estimated:

Cation Resin	Anion Resin
TC as $CaCO_3$ = 112 mg/L 112/17.1 = 6.6 g/gal %Na $= \dfrac{55 \text{ mg/L} \times 100}{112 \text{ mg/L}} = 49\%$	TEA as $CaCO_3$ = CO_2 5 mg/L $\times$ 1.13 = 6 mg/L SiO_2 7.4 $\times$ 0.83 = 6 mg/L $\rightarrow$ Sulfates = 21 mg/L $\rightarrow$

Cation Resin	Anion Resin

Cation Resin

$$\%\text{alk.} = \frac{69 \text{ mg/L} \times 100}{112 \text{ mg/L}} = 62\%$$

Since the cation leakage is $= 22\%$

$$\frac{2 \text{ mg/L} \times 100}{112 \text{ mg/L}} = 1.8\%$$

We should select a strong acid cation exchanger, and from Table 7-4 use a regenerant dosage of 5 lb H_2SO_4 per ft^3 will yield capacity of 13.6 kg/ft^3 of bed TC = Total cations

Anion Resin

$$\text{Chlorides} = \frac{22 \text{ mg/L} \rightarrow}{-55 \text{ mg/L} \rightarrow}$$

$$55/17.1 = 3.2 \text{ g/gal}$$
$$\%\text{Weak acids} = 12 \times 100/55$$

$\%$Chlorides $= 22 \times 100/55 = 40\%$
$\%$Silica $= 6 \times 100/55 = 11\%$
Since the required silica leakage must be less than 0.03 mg/L, we should select a type-I porous strong base anion exchanger. From Table 7-6 a regenerant dosage of 5 lb NaOH per ft^3 will yield a capacity of 10.6 kg/ft^3 of bed.

$$\text{TEA} = \text{Total exchangeable anions}$$

(c) Design of Anion Units. The area of the anion units and the resin volume required are first calculated. Since a continuous flow of 266 gal/min is required without demineralized water storage, we will use two anion units.

At a flowrate of 12 gal/min/ft^2 through one unit while the other is being regenerated, we will need a bed area of:

$$\frac{266 \text{ gal/min}}{12 \text{ gal/min/ft}^2} = 22.2 \text{ ft}^2.$$

This can be accomplished with a 5' 5" diameter vessel (23.5 ft^2). Assuming two regenerations per day, the estimated bed depth is:

$$\frac{266 \text{ gal/min} \times 1440 \text{ min} \times 3.2 \text{ g/gal} \times 29 \text{ ft}^3}{2 \text{ U} \times 2 \text{ regens} \times 10.6 \text{ kg/ft}^3 \times 23.5 \text{ ft}^2} = 1.2 \text{ ft}$$

This is much too shallow. Using a minimum depth of 2.5 ft, we will have 2.5 $\times$ 23.5 = 59 ft^3 in each anion unit, and the straight side shell height, with 100% freeboard, is 2 $\times$ 2.5 = 5 ft. So that we now have two units each 5' 6" in diameter by 5 ft high and containing 59 ft^3 of a type-I strong base anion exchanger with a total operating capacity of: 59 ft^3 $\times$ 10.6 kg/ft^3 = 625 kg.

The run length can now be estimated as follows:

$$\frac{625 \text{ kg} \times 1000}{3.2 \text{ g/gal} \times 133 \text{ gal/min} \times 60 \text{ min}} = 24.4 \text{ hr}$$

Therefore, each anion unit will be regenerated once each day instead of twice.

At 5 lb/ft^3 the caustic soda requirement is 295 lb per regeneration, or 1.5 lb of 100% caustic per 1000 gal of water treated.

(d) Anion Regeneration Schedule. A 4% solution of caustic at 120°F will be used:

4% caustic contains 0.348 lb NaOH/gal
50% caustic contains 6.364 lb NaOH/gal

The steps required in the regeneration are:

1. Backwash: 3 gal/min/ft² × 15 min × 23.5 ft² = 1070 gal
2. Caustic Injection: Total NaOH required (100% basis) will be

$$a = \frac{295 \text{ lb}}{6.364 \text{ lb/gal (50\%)}} = 46 \text{ gal of 50\% NaOH}$$

$$b = \frac{295 \text{ lb}}{0.348 \text{ lb/gal (4\%)}} = 848 \text{ gal of 4\% NaOH}$$

Therefore, the dilution water required is 848 − 46 = 802 gal in about 60 min, and this will be taken from the cation exchanger effluent. The regenerant rate should be 0.25 gal/min/ft³ × 59 ft³ = 15 gal/min, and will require about 848 gal/(15 gal/min) = 59 min.

The dilution water flowrate is 802 gal/56 min = 14.3 gal/min, and the 50% NaOH flowrate will be 46 gal/56 min = 0.82 gal/min.

3. Displacement (Slow) Rinse: The slow rinse flowrate is the same as the dilution flowrate, or 14.3 gal/min. The rinse volume is determined by the void volume of the bed (40%), and the 6-in space between the top of the bed and the regenerant distributor.

$$\left[\left(\frac{40\%}{100} \times 59 \text{ ft}^3 \right) + (0.5 \text{ ft} \times 23.5 \text{ ft}^2) \right] \times 7.5 = 265 \text{ gal}$$

so that the time required is 256 gal/14.3 gal/min = 18.5 min.

4. Rinse (Fast) to Waste: The flowrate of the fast rinse can be 6 gal/min/ft², or 141 gal/min. The volume used is usually 100 gal/ft³; and, therefore, the total fast rinse volume will be 5900 gal. About one half of this rinse, or 2950 gal, can be recycled back to the cation inlet. The cation effluent is used for backwashing and rinsing the anion exchange unit. In addition, the stored demineralized water is used to dilute the caustic, and for the displacement (slow) rinse.

(e) Design of Cation Units. The cation unit must provide the anion regeneration water. The recycled anion rinse will contain only one half of the total dissolved solids of the raw water; therefore, we can assume that 1475 gal needs to be included. So that the total process volume required is:

Anion regenerant water	5087 gal
Recycled rinse	1475 gal
Service water	96,000 gal
Total	102,562 gal

(102,562 gal × 6.6 g/gal)/1000 = 678 kg.
678 kg/13.6 kg/ft^3 = 50 ft^3 of resin bed.)

Therefore, we will use two cation units each 5' 6" in diameter and 5' high, containing 59 ft^3 with a total capacity of 800 kg. This cation unit will yield a process volume of about 121,000 gallons between regenerations. The length of the cation run can now be estimated as follows: (121,000 − 6562)/8000 gal/hr = 14.3 hr, instead of the 12 hr originally assumed.

The total acid required = 59 ft^3 × 5 lb/ft^3 = 295 lb, or 295/114 = 2.67 lb H$_2$SO$_4$ per 1000 gal of product water.

(f) Cation Regeneration Schedule.

1. Backwash: 6 gal/min/ft^2 × 23.5 ft^2 × 15 min = 2120 gal at a flowrate of 141 gal/min.

2. Sulfuric acid injection: This regeneration will be carried out in two steps, using 2% and 4% to avoid as much as possible the precipitation of calcium sulfate in the resin bed. Since the Ca/TC = 0.31, we will use a dilute acid rate of 31/50 = 0.6 gal/min/ft^3.

   $$0.6 \text{ gal min/ft}^3 \times 59 \text{ ft}^3 = 35.4 \text{ gal/min}$$

   $$2\% \text{ acid} = 0.17 \text{ lb of H}_2\text{SO}_4/\text{gal}$$

   $$4\% \text{ acid} = 0.34 \text{ lb of H}_2\text{SO}_4/\text{gal}$$

   $$66° \text{ Baumé acid} = 15.3 \text{ lb of H}_2\text{SO}_4/\text{gal}$$

Assuming that one half of the acid injection will be at 2% and the remainder at 4%:

For 2% acid:

$$a = \frac{295/2}{0.17 \text{ lb/gal}} = 870 \text{ gal of 2\% acid}$$

$$b = \frac{295/2}{15.3 \text{ lb/gal}} = 9.6 \text{ gal of 66° Be H}_2\text{SO}$$

And the volume of dilution water is equal to $a - b$ or 860 gal. The dilution water rate is 774/24.5 min = 35.1 gal/min.

For 4% acid

$$a = \frac{295/2}{0.34 \text{ lb/gal}} = 433 \text{ gal}$$

$$b = \text{strong acid} = 96 \text{ gal}$$

The volume of dilution water required is $a - b = 337$ gal, since the dilution flow is fixed at 31.6 gal/min. The time required for the 4% injection is $337/31.6 = 10.7$ min.

3. Displacement (slow) rinse: The volume of this rinse is the same as the anion slow rinse; however, since the flowrate is higher, the time required is reduced to 8.4 min.
4. Fast rinse: The fast rinse is estimated as follows

$$\frac{50 \text{ gal/min/ft}^3 \times 59 \text{ ft}^3}{141 \text{ gal/min}} = 21 \text{ min}$$

(g) Simultaneous Regeneration of Anion and Cation Units. The total outage time for regeneration of the anion and cation exchangers separately, as shown in Tables 7-10 and 7-11 is 211.1 min. This can be reduced by regenerating these units simultaneously, as indicated in Table 7-12. Thus, the cation regeneration is completed while the caustic is flowing through the anion unit, and a saving of 79.6 min is accomplished.

(h) Vacuum Deaerator. The vacuum deaerator recommended for this application will be a rubber-lined steel shell, 4′ 6″ in diameter by 9′ tall, using rotary vacuum pumps.

(i) Regenerant Costs. Assuming that the cost of sulfuric acid (100% basis) is 1 cent per pound, and that of caustic (100% basis) is 3 cents per pound, the regenerant costs can be calculated as follows:

$$\text{Acid: } 2.6 \text{ lb} \times 1 \text{ cent/lb} = 2.6$$

$$\text{Caustic: } 1.5 \text{ lb} \times 3 \text{ cents/lb} = \underline{4.5}$$

$$7.1 \text{ cents/1000 gal}$$

Table 7-10.   Case A: Summary of Anion Regeneration Schedule

Operation	Flowrate (gal/min)	Duration (min)	Waste (gal)	Recycle (gal)
Backwash	71	15	1070	
Dilution	14.3	56	802	
Displacement	14.3	18.5	265	
Fast rinse	141	21	2950	
Rinse recycle	141	21		2950
Totals		131.5	5087	

Table 7-11.   Case A: Summary of Cation Regeneration Schedule

Operation	Flowrate (gal/min)	Duration (min)	Volume (gal)
Backwash	141	15	2120
2% acid injection	31.6	24.5	860
4% acid injection	31.6	10.7	3.37
Displacement rinse	31.6	8.4	265
Fast rinse	141	21	2950
Totals		79.6	6198

(j) Design of Pretreatment Equipment. The total flow required is estimated as follows:

Final effluent flowrate: 266

Anion regeneration, 5087 gal/U/day

$$\frac{1 \times 2 \text{ U } 5087 \text{ gal}}{1440 \text{ min/day}} = \quad 7.0$$

Cation regeneration, 6446 gal/U/day, twice each day

$$\frac{2 \times 2 \text{ U } 6446 \text{ gal}}{1440 \text{ min/day}} = \quad 17.9$$

$$\frac{\text{Filter backwash to waste } 26}{\text{Total } 317 \text{ gal/min}}$$

The coagulator and two gravity filters are designed to provide a total flowrate of 350 gal/min. The clear, well below the filters, provides storage capacity to satisfy filter backwash, cation regeneration, and rinse requirements.

The coagulator is designed for a rise rate of 1.1 gal/min/ft^2, and a unit 25' 6" in diameter by 13' 6" in height will be required.

Table 7-12.   Case A: Simultaneous Regeneration

Anion Units		Cation Unit	
Process	Min	Process	Min
Wait		Backwash	15
Backwash	15	Caustic	56
2% acid	24.5		
4% acid	10.7		
Slow rinse	8.4		
Fast rinse	21	Slow rinse	18.5
	79.6		89.5
Wait	9.9	Fast rinse	42
Total min	89.5		131.5

The gravity filters are designed for a rate of 2.2 gal/min/ft^2 and two will be necessary, each measuring 12' 6" in diameter by 12 ft high.

Chemical feed equipment of the solution type will be required for chlorine; lime, ferric sulfate and coagulant are needed for the pretreatment process.

Case B: Chlorination and Coagulation Pretreatment Followed by Primary Two-Bed Demineralization, Deaeration, and a Secondary Mixed-Bed Polishing Unit (Figure 7-2).

1 Design Data.   The total flow required is 1000 gal/min without demineralized water storage for make-up to a once-through supercritical boiler (3500 psi).

The water supply is from a river high in color, turbidity, and organic matter. The maximum and average water analysis is shown in the first two columns of Table 7-13.

The final treated water quality requirements are:

$$\text{Total dissolved solids} < 0.4 \text{ mg/L}$$

$$\text{Conductivity} < 1.0 \text{ micromho}$$

$$\text{Residual silica} < 0.01 \text{ mg/L as SiO}$$

2 Modification of Water Analysis by Pretreatment.   The dosages of chemicals in mg/L for this application are:

	Average	Maximum
Chlorine	5	10
Aluminum sulfate	34	51
Hydrated lime (93%)	9	16

The high dosages of chlorine and aluminum sulfate assumed are necessary to reduce the turbidity, color, and organic matter contained in this water supply. The lime dosage was selected to produce a final pH of 6.5, which should be optimum for the coagulation process.

The raw water average and maximum analysis is modified as shown for Case 1, and the adjusted analytical results are given in Table 7-13. The *design analysis* used in the calculation will be the average after modification for the pretreatment.

3 Design Calculations.   (a) Definition of Demineralizer System. The "very pure" effluent quality requirements dictates the use of a multibed treatment system with primary units followed by secondary mixed-bed units. The mixed-bed is selected because it will produce the best effluent water quality. The use of primary units will reduce over all operating costs and will protect the strong base anion resin in the secondary units from organic fouling. Since the flow for

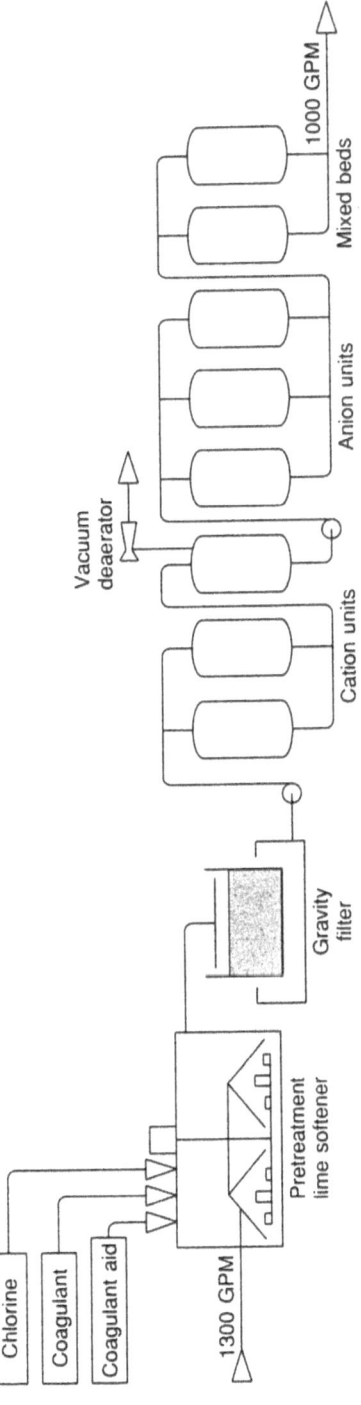

Figure 7-2. Case B: Final process schematic.

Table 7-13.   Case B: Estimated Operating Performance[a,b]

	A	B	C	D	E	F	G
Calcium	150	100	112	0	0	0	0
Magnesium	75	50	50	0	0	0	0
Sodium	150	100	100	20	20	20	0.4
Hydrogen	0	0	0	212	212	0	0
Total Cations	375	250	262	232	232	20	0.4
Bicarbonate	50	40	30	0	0	0	0
Carbonate	0	0	0	0	0	0	0
Hydroxyl	0	0	0	0	0	10	0.4
Sulfate	175	120	135	135	135	0	0
Chloride	150	90	97	97	97	10	0
Total Anions	375	250	262	232	232	20	0.4
Silica as SiO_2	30	20	20	20	20	0.1	0.01
CO_2 as CO_2	10	10	18	204	10	0	0
Turbidity JTU^c	400	100	0	0	0	0	0
Color	80	40	5	5	5	5	5
Organics[d]	40	30	2	0	0	0	0
pH	7.0	6.9	6.5	3.0	6.5	9.0	7.0

[a]A, maximum raw water analysis; B, average raw water analysis; C, average analysis after pretreatment (design analysis); D, effluent from cation exchanger; E, after vacuum deaerator; F, effluent from anion exchanger; G, final treated water (mixed-bed effluent).
[b]All values are mg/L as $CaCO_3$ unless shown otherwise.
[c]JTU, Jackson turbidity units.
[d]Based on titration with $KMnO$.

this application is high, 1000 gal/min, a vacuum deaerator is included to reduce the free carbon dioxide found in the cation effluent; this will reduce the amount of caustic needed to regenerate the primary anion exchanger.

(b) Selection of Resins, Regeneration Dosages, and Capacities for the Average Pretreatment Water Quality.

Primary units

Cation Resin	Anion Resin
$TC = \dfrac{262 \text{ mg/L}}{17.1} = 15.4 \text{ g/gal}$	TEA after deaeration
$\%Na = \dfrac{100 \text{ mg/L} \times 100}{262 \text{ mg/L}} = 38\%$	$CO_2 = 5 \text{ mg/L} \times 1.13 = \quad 6 \text{ mg/L}$
	$SiO = 20 \text{ mg/L} \times 0.83 = \quad 17 \text{ mg/L}$
	$SO_4 = \qquad\qquad 135 \text{ mg/L}$
$\%Alk. = \dfrac{30 \text{ mg/L} \times 30}{262 \text{ mg/L}} = 11.5\%$	$Cl = \qquad\qquad 97 \text{mg/L}$
	255 mg/L
In order to obtain a leakage of 0.4 mg/L from the secondary units, it will be necessary to reduce the primary effluent to about 20	$\dfrac{255 \text{ mg/L}}{17.1} = 14.9 \text{ g/gal}$

Cation Resin	Anion Resin
mg/L TDS. Therefore, the primary design leakage is:	$$\% \text{ SiO} = \frac{17 \times 100}{255 \text{ mg/L}} = 6.7\%$$

<table>
<tr><td>

$$\frac{20 \text{ mg/L} \times 100}{262 \text{ mg/L}} = 7.6\%$$

This can be realized with a dosage of 4.0 lb sulfuric acid per ft^3 on a strong acid cation exchanger. This dosage will yield an operating capacity of 11.0 kg ft^3.

</td><td>

$$\% \text{Weak acids} \frac{23 \text{ mg/L} \times 100}{255 \text{ mg/L}} = 9\%$$

$$\%\text{Cl} = \frac{9.7 \times 100}{255 \text{ mg/L}} = 38\%.$$

Assume that the primary SiO leakage will be about 0.1 mg/L. We will use a type-I strong base anion exchanger, which will yield a capacity anion exchanger, which will yield a capacity of 9.5 kg/ft^3 at a regenerant dosage of 4 lb NaOH/ft^3.

</td></tr>
</table>

Secondary Mixed Beds

Cation Resin	Anion Resin
$$\text{TC} = \frac{20 \text{ mg/L}}{17.1} = 1.2 \text{ g/gal}$$	$$\text{TEA} = \frac{20.08 \text{ mg/L}}{17.1} = 1.2 \text{ g/gal}$$

<table>
<tr><td>

%Na = 100%

%Alk. = 50%

However, for mixed beds the cation leakage and capacity are always based on 100% alkalinity, and the allowable cation leakage is:

$$\frac{0.4 \text{ mg/L} \times 100}{20 \text{ mg/L}} = 2\%$$

Therefore, use a strong acid cation exchanger. A dosage of 5 lb ft^3 of sulfuric acid will yield an operating capacity of: 21.4 × 0.8 = 17.1 kg/ft^3.

</td><td>

% weak acids = nearly 0

% SiO$_2$ = nearly 0

Use a porous type-I strong base anion exchanger and a regenerant dosage of 6 lb/ft^3 NaOH. Correcting for the Cl ion present will yield a capacity of 10.2 kg/ft^3 and a final operating capacity of: 10.2 × 0.8 = 8.2 kg/ft^3.

</td></tr>
</table>

In the above table, the factor of 0.8 is an adjustment for the exchange efficiency when these ion exchangers are used in a mixed bed; TC = total cations; TEA = total exchangeable anions.

(c) Design of Mixed Beds. At a flowrate of 15 gal/min/ft^2, we will need:

$$\frac{1000 \text{ gal/min}}{15 \text{ gal/min/ft}^2} = 63 \text{ ft}^2, \text{ or } 9' \ 0'' \text{ in diameter}$$

For 2 units regenerated once every 3 days, the process volume per unit is:

$$\frac{1,440,000 \text{ gal/day} \times 3 \text{ days}}{2 \text{ U}} = 2,160,000 \text{ gal/U}$$

The bed depth of each component in the mixed bed can now be estimated as follows:

$$\text{Cation resin} \frac{2,160,000 \text{ gal} \times 1.2 \text{ g/gal}}{17.1 \text{ kg/ft}^3 \times 63 \text{ ft}^2} = 2.4 \text{ ft depth}$$

$$\text{Anion resin} \frac{2,160,000 \text{ gal} \times 1.2 \text{ g/gal}}{8.2 \text{ kg/ft}^3 \times 63 \text{ ft}^2} = 5.0 \text{ ft depth}$$

and the straight side shell dimensions are as follows:

Cation portion	2.4 ft
Anion portion	5.0 ft
100% freeboard	7.4 ft
Shell height	14.8 ft

Using 2 units 9 ft in diameter by 15 ft deep, we can calculate the regenerant usage for this application:

$$H_2SO_4 = \frac{5 \text{ lb/ft}^3 \times 151 \text{ ft}^3}{2160 \text{ M gal}} = 0.4 \text{ lb/1000 gal}$$

$$NaOH = \frac{6 \text{ lb/ft}^3 \times 316 \text{ ft}^3}{2160 \text{ M gal}} = 0.9 \text{ lb/1000 gal}$$

(d) Regeneration Schedule for Secondary Mixed Beds. Simultaneous regeneration of the anion and cation portions of these mixed beds can be accomplished without danger of precipitating calcium sulfate, since there is little calcium ion on the exhausted mixed bed. The calculations for each step of the simultaneous regeneration are as follows:

1. Backwash: The backwash is carried out for 15 min at a flowrate of 3 gal/min/ft² at 70°. The low flowrate is used to avoid losing any anion resin during the backwash step. The backwash volume is:

$$9 \text{ ft} \times 63.5 \text{ ft}^2 \times 3 \text{ gal/min/ft}^2 \times 15 \text{ min} = 25718 \text{ gal}$$

Clean water should be used for the backwash to avoid clogging the underdrain screens.

2. Simultaneous regeneration: Simultaneous regeneration is accomplished by using 4% sulfuric acid for the cation resin portion, and 4% caustic soda for the anion portion.

For the acid: 4% contains 0.34 lb H_2SO_4 per gal; 66° Baume contains 15.3 lb H_2SO_4 per gal; and the H_2SO_4 required is 755 lb per regeneration:

$$4\% \text{ acid} = \frac{755 \text{ lb}}{0.34 \text{ lb/gal}} = 2220 \text{ gal}$$

$$\text{Strong acid} = \frac{755 \text{ lb}}{15.3 \text{ lb/gal}} = 49 \text{ gal}$$

$$\text{Dilution water} = \overline{2171 \text{ gal}}$$

For the NaOH: 4% contains 0.348 lb NaOH gal; 50% contains 6.364 lb NaOH/gal; NaOH required is 1896 lb per regeneration:

$$4\% \text{ NaOH} = \frac{1896 \text{ lb}}{0.348 \text{ lb/gal}} = 5448 \text{ gal}$$

$$50\% \text{ NaOH} = \frac{1896 \text{ lb}}{6.364 \text{ lb/gal}} = 298 \text{ gal}$$

$$\text{Dilution water} = \overline{5190 \text{ gal}}$$

With a 60 min contact time:

$$H_2SO_4 \text{ dilution rate} = \frac{2171 \text{ gal}}{60 \text{ min}} = 36 \text{ gal/min}$$

$$\text{NaOH dilution rate} = \frac{5150 \text{ gal}}{60 \text{ min}} = 86 \text{ gal/min}$$

3. Displacement (slow) rinse: The displacement rinse is carried out with the same dilution flows calculated in step 2 above, the total volume of rinse water produced in this step is:

For the anion portion

```
Voids  = 316 ft³ × 40% × 7.5 gal/ft³              =  948 gal
Water  = 0.5 ft × 63.5 ft² × 7.5 gal/ft³          =  240 gal
                                      Subtotal     = 1188 gal
       Time = 1188 gal/86 gal/min = 13.8 min.
         For the cation portion 13.8 min × 36 gal/min  =  497 gal
Total rinse to waste                               = 1685 gal
```

4. Drain and air mix: Draining the excess water to the top of the bed will take about 15 min. Air-mixing will require about 10 std. ft³/min/ft²

of bed area; therefore, the air flow needed for this bed is:

$$63.5 \text{ ft}^2 \times 10 \text{ ft}^3/\text{min} = 635 \text{ ft}^3/\text{min at about 10 psi}$$

5. Refill: Refilling the voids after air-mixing will require two or three steps. The volume needed is given in Table 7-14.
6. Final (fast) rinse: The fast rinse uses 25 gal/ft^3 of bed, at a flowrate of 6 gal/min/ft^2, and the rinse volume is:

$$(151 + 316) \text{ ft}^3 \times 25 \text{ gal/ft}^3 = 11,700 \text{ gals.}$$

$$\text{Time} = \frac{11,700 \text{ gals}}{6 \text{ gal/min}} = 31 \text{ min}$$

A summary of the regeneration steps for the secondary mixed beds is given in Table 7-14.

(e) Design of Primary Units. Since the secondary mixed beds are regenerated only once every 3 days, their outage periods can be selected to come when the primary units are in operation. As a result, it is usual to neglect the mixed-bed water requirement when designing the primary units. However, the primary cation unit capacity must include the water required for the regeneration of the primary anion units. For this purpose, 100–130 gal/ft^3 of primary anion resin may be assumed.

Table 7-14.   Case B: Summary of Mixed-Bed Regeneration Schedule

Operation	Cation Portion			Anion Portion		
	Flow (gal/min)	Duration (min)	Total (gal)	Flow (gal/min)	Duration (min)	Total (gal)
Backwash		15		190	15	2850
Regeneration	36	60	2171	86	60	5182
Slow rinse	36	13.8	495	86	13.8	1188
Air mixing						
Drain				—	15	
Refill				190	2	380
Air mix				635^a	10	
Drain				(a)	5	
Air mix				635^a	5	
Fill slow				40	10	400
fast				320	10	3200
slow				40	5	200
Fast rinse				380	31	11,700
	Totals		2266	2266	181.8	25,100

^{a}Cu. ft/min.

Primary Units

Anion Units	Cation Units

Anion Units

Flow will be 10 gal/min/ft² with 2 U on stream, while the third is being regenerated:

$$\frac{1000 \text{ gal/min}}{10 \text{ gal/min/ft}^2} = 100 \text{ ft}^2$$

Use 3 U 8 ft in diameter each with 50 ft³. Use 12-hr runs at the average water analysis of 255 mg/L; then at a maximum of 393 mg/L the run time will be:

$$\frac{255 \text{ mg/L}}{393 \text{ mg/L}} \times 12 = 7.8 \text{ hr}$$

This will not be too short, since the anion regeneration takes 3 hr. The run length must exceed the regenerant time multiplied by the number of units minus 1:

$$(3 - 1) \times 3 \text{ hr} = 6 \text{ hr.}$$

which is less than the 7.8 hr with the maximum water analysis. The service run therefore is:

$$\frac{60 \text{ M gal} \times 12 \text{ hr}}{3 \text{ U}} = \frac{240 \text{ M gal}}{\text{U}}$$

240 M gal × 14.9 g/gal = 3580 kg
Use a type-I strong base anion exchanger with a bed volume of:

$$\frac{3580 \text{ kg}}{9.5 \text{ kg/ft}^3} = 377 \text{ ft}^3$$

$$\text{Depth} = \frac{377 \text{ ft}^3}{50 \text{ ft}^2} = 7.5 \text{ ft}$$

$$100\% \text{ Freeboard} = \underline{7.5 \text{ ft}}$$

$$\text{Straight height} = 15.0 \text{ ft}$$

We will use 3 units 8 ft in diameter by 15 ft high, with 377 ft³ of resin in each.
NaOH = 4 lb × 377 ft³ = 1508 lb
$$\frac{1508 \text{ lb}}{240 \text{ M gal}} = 6.3 \text{ lb NaOH per 1000 gal}$$
    of service water

Cation Units

Also, use 10 gal/min/ft²

Same as anion units

Service run	240 M gal
Anion Regen.	
377 × 130 gal	49 M gal
Total	289 M gal

289 M gal × 15.4 g/gal = 4450 kg
Use a strong acid cation exchanger with a bed volume of:

$$\frac{4450 \text{ kg}}{11 \text{ kg/ft}^3} = 405 \text{ ft}^3$$

$$\text{Depth} = \frac{405 \text{ ft}^3}{50 \text{ ft}^2} = 8.1 \text{ ft}$$

$$75\% \text{ freeboard} = \underline{6.0 \text{ ft}}$$

$$\text{Straight height} = 14.1 \text{ ft}$$

We will use 3 units 8 ft in diameter by 14 ft high, with 405 ft³ of resin in each.
Acid = 4 lb × 405 ft³ = 1620 lb
$$\frac{1620 \text{ lb}}{240 \text{ M gal}} = 6.7 \text{ lb acid per 1000 gal}$$
    of service water

(f) Design of Vacuum Deaerator and Transfer Pumps. Omitting the vacuum deaerator would increase the TEA to the anion exchange units by almost 20%. This would increase the operating cost at least $26,000 per year; therefore, the deaerator is economically justified for this application. It will reduce the carbon dioxide to about 5 mg/L and the dissolved oxygen to less than 1 mg/L.

The flow through the vacuum deaerator and the transfer pumps is as follows

$$\text{Primary anion rinse at } 50 \text{ ft}^2 \times 6 \text{ gal/min/ft}^2 = 300 \text{ gal/min}$$

$$\text{Service flow} = 1000 \text{ gal/min}$$

$$\text{Total} = 1300 \text{ gal/min}$$

The deaerator will be a rubber-lined shell 7 ft in diameter by 12 ft high, with Raschig-ring fill and an evacuator using a steam jet.

The transfer pumps should be in duplicate and made of stainless steel (one being a spare or stand-by). Each pump will have a capacity of 1300 gal/min; alternately three pumps could be installed, each having a capacity of 650 gal/min. The effective head of the pumps must be sufficient to overcome the 15 pressure losses which occur through the anion and downstream mixed-bed units in line and still provide a final pressure head for the treated effluent to reach the final deaerating heater.

4 Design of Pretreatment Plant. The pretreatment plant is designed in the same way as in Case A; Therefore, the detailed calculations are not included here.

Case C: Chlorination and Coagulation Pretreatment Followed by Primary Three-Bed and Secondary Mixed-Bed Demineralization Without Deaeration (Figure 7-3)

1 Design Data. The design data are the same as given for Case B.

2 Modification of Water Analysis by Pretreatment. The modifications to the water analysis after the addition of pretreatment chemicals is the same as given in Case B, and is shown in detail in Table 7-13 (see column C). The estimated performance of this treatment system is shown in Table 7-15.

3 Design Calculations. (a) Design of Demineralizer System. In this application, a three-bed primary system is used for reducing the quantity of caustic soda needed for regeneration. The vacuum deaerator has been eliminated, since the strong base resins have excess capacities, and they can remove the free carbon dioxide economically. The primary weak base resin and the strong acid cation resins are regenerated at the same time, when the weak base resin effluent

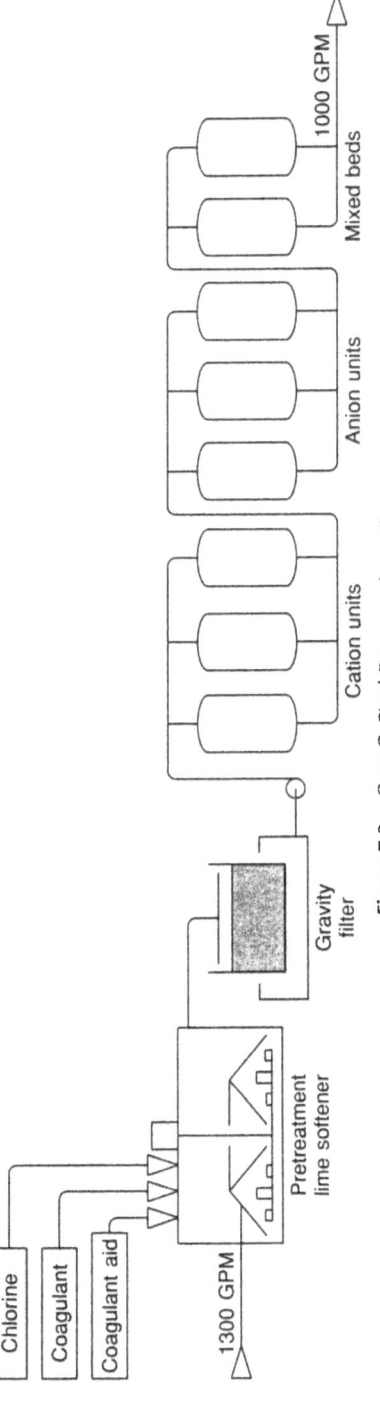

Figure 7-3. Case C: Final Process schematic.

Table 7-15.   Case C: Estimated Operating Performance[a,b]

	A (max)	B (min)	C (aver.)	D	E	F	G
Calcium	150	100	112	0	0	0	0
Magnesium	75	50	50	0	0	0	0
Sodium	150	100	100	20	20	20	0.4
Hydrogen	0	0	0	212	0	0	0
Total cations	375	250	262	232	20	20	0.4
Bicarbonate	50	30	40	0	0	0	0
Carbonate	0	0	0	0	0	0	0
Hydroxyl	0	0	0	0	0	10	0.4
Sulfate	175	120	135	135	0	0	0
Chloride	150	90	97	97	20	0	0
Total anions	375	250	262	232	20	10	0.4
CO_2 as CO_2	10	10	18	204	10	0	
Silica as SiO_2	30	20	20	20	20	0.1	0.01
Turbidity JTU[c]	400	100	0	0	0	0	0
Color	80	40	5	5	5	5	5
Organics[d]	40	3	2	0	0	0	0
pH	7.0	6.9	6.5	3.0	6.0	9.0	7.0

[a]A, maximum raw water analysis; B, average raw water analysis; C, average analysis after pretreatment (design analysis); D, effluent from cation exchanger; E, effluent from weak base anion exchanger; F, effluent from strong base anion exchanger; G, final treated water (mixed-bed effluent).
[b]All values are mg/L as $CaCO_3$ unless shown otherwise.
[c]JTU, Jackson turbidity units.
[d]Based on titration with KMnO.

shows a conductivity increase (breakthrough). The caustic soda regenerant goes through the system countercurrently, first through the strong base resin bed and then through the weak base resin bed.

(b) Selection of Resins, Regeneration Dosages, and Capacities. The primary cation units are the same as in Case B. The strong acid cation exchanger will yield a capacity of 11.0 kg/ft³ at a regenerant dosage of 4 lb of acid/ft³. The regenerant dosages and capacities of the primary weak base and strong base anion resins are calculated as follows:

Weak Base Resin	Strong Base Resin
Sulfate = 135 mg/L	$CO_2 = 44 \times 1.13 = 51$ mg/L
Chloride = 97 mg/L	$SiO_2 = 20 \times 0.83 = 17$ mg/L
TMA = 232 mg/L	Cation leakage = 20 mg/L
$TMA = \dfrac{232 \text{ mg/L}}{17.1} = 13.6$ g/gal	TEA = 88 mg/L
$\%Cl = \dfrac{97 \times 100}{232} = 42\%$	$TEA = \dfrac{88 \text{ mg/L}}{17.1} = 5.2$ g/gal
	$\%\text{weak acids} = \dfrac{68 \times 100}{88} = 78\%$

Weak Base Resin	Strong Base Resin
	$\%SiO_2 = \dfrac{17 \times 100}{88} = 20\%$
	$\%Cl = \dfrac{10 \times 100}{88} = 11.4\%$
Use a weak base anion resin which will yield a capacity of 19 kg/ft^3 at a dosage of 4 lb of NaOH per ft^3.	Use a porous, type-I strong anion resin which will yield a capacity of 13.5 kg/ft^3 at a dosage of 5 lb of NaOH per ft^3.

In the above table, TMA = total mineral acidity, and TEA = total exchangeable anions.

Note that the cation leakage is included in the TEA for the primary strong base anion exchanger; this is necessary because the strong base resin will convert salts to their corresponding bases, and this process uses a portion of the available capacity.

The resins, regenerant levels, and capacities for the secondary mixed beds are the same as given in Case B.

(c) Design of Mixed Beds. See Case B.

(d) Design of Primary Units. The primary cation units are the same as Case B. The primary weak base unit is designed as follows.

Using a flowrate of 10 gal/min/ft^2 with 1 U off-line, the required bed area is:

$$\frac{1000 \text{ gal/min}}{10 \text{ gal/min/ft}^2} = 100 \text{ ft}^2$$

Use three units, each 8 ft in diameter with 50 ft^3 per unit.

Use 12-hr runs, based on the average water analysis given in Table 7-13 in Case B. Then the estimated total capacity required is:

$$\frac{1000 \text{ gal/min} \times 60 \text{ min} \times 12 \text{ hr}}{3 \text{ U} \times 1000 \text{ g/kg}} \times 13.6 \text{ g/gal} = 3270 \text{ kg}$$

Since the resin capacity is 19 kg/ft^3, we will need:

$$\frac{3270 \text{ kg}}{19 \text{ kg/ft}^3} = 172 \text{ ft}^3 \text{ of weak base resin per unit}$$

$$\text{Weak base depth} = \frac{172 \text{ ft}^3}{50 \text{ ft}^2} = 3.5 \text{ ft}$$

20% swelling	= 0.7 ft
100% freeboard	= 4.2 ft
Straight shell height	= 8.4 ft

Therefore, use 3 units each 8' in diameter and 8'6" high, containing 172 ft^3 of weak base resin. At 4 lb of NaOH/ft^3 this will require $172 \times 4 = 688$ lb of NaOH per regeneration.

To this must be added the kilograin equivalent of the caustic soda consumed in the regeneration of the strong base anion resin. The caustic soda passing through the strong base resin is converted to sodium silicate and sodium carbonate. While these weak alkalies can regenerate the weak base resin, it is usual to assume that the caustic soda taken up by the strong base resin should be included in the total caustic soda used for the countercurrent regeneration.

The strong base anion units are designed as follows:

$$\frac{240,000 \text{ gal}}{1000 \text{ g/kg}} \times 5.2 \text{ g/gal} = 1248 \text{ kg}$$

$$\frac{1248 \text{ kg}}{13.5 \text{ kg/ft}^3 \times 50 \text{ ft}^2} = 1.8 \text{ ft bed depth}$$

This is too shallow; using a recommended bed depth of 2.5 ft, each unit will contain:

$$50 \text{ ft}^2 \times 2.5 \text{ ft} = 125 \text{ ft}^3.$$

Therefore, use three units 8 ft in diameter by 5 ft deep, filled with a porous type-I strong base anion resin.

The strong base resin will only be 73% exhausted at the end of the run.

$$\text{NaOH equivalent of 1248 kg} = \frac{1248}{7} \times 0.8 = 143 \text{ lb}$$

$$\text{NaOH for the weak base resin} = \underline{688 \text{ lb}}$$

$$\text{Total NaOH} \qquad\qquad\qquad 831 \text{ lb}$$

The 831 lb of caustic soda yields a dosage of:

$$\frac{831 \text{ lb}}{125 \text{ ft}^3} = 6.6 \text{ lb NaOH/ft}^3 \text{ of bed}$$

This is well above the 5 lb/ft^3 previously specified. The NaOH consumed per 1000 gal is:

$$\frac{831 \text{ lb NaOH}}{240 \text{ M gal}} = 3.5 \text{ lb/1000 gal}$$

(e) Regenerant Costs. Assuming the cost of sulfuric acid to be 1 cent/lb, and that caustic soda cost is 3 cents/lb (both on a 100% basis), we can calculate the total regenerant costs as follows:

	lb	Cents	Total Cost
Primary acid	6.7		
Secondary acid	0.4		
Total acid	7.1	1.0	7.1
Primary caustic soda	3.5		
Secondary caustic soda	0.9		
Total caustic soda	4.4	3.0	13.2
Total regenerant cost			20.3

(g) Regenerant Cost for Comparison. A comparison of the regenerant costs for cases B and C can be made in the following way:

$$\text{Cents}/1000 \text{ gal}$$

Regenerant costs Case B	28.7
Regenerant costs Case C	20.3
Difference (savings)	$8.4/1000$ gal

This saving is due to the use of the three-bed system, instead of the two-bed system, in the primary stage, and obtains even though the vacuum deaerator is omitted from Case C.

Excess acid:

$$\text{lb acid per regeneration per unit} \quad 1620$$

$$\text{acid consumed } \frac{4450 \text{ kg}}{7} = \underline{636}$$

$$\text{net excess} \qquad\qquad 984 \text{ lb}$$

Excess caustic soda:

$$\text{lb NaOH per regeneration per unit} \quad 831$$

$$\text{expressed as CaCO}_3 \frac{831}{0.8} = 1040$$

$$\text{NaOH consumed } \frac{3270 \text{ kg}}{7} = \underline{467}$$

$$\text{net excess} \qquad\qquad 573 \text{ lb}$$

So that

$$\text{Excess acid} = 984 \text{ lb}$$

$$\text{Excess NaOH} = \underline{573 \text{ lb}}$$

$$\text{Net excess acid} = 411 \text{ lb}$$

If this excess acid cannot be used for some plant application, it will have to be neutralized with additional caustic soda before being discharged. Therefore, the apparent savings in caustic soda by using the three-bed system is not realized.

Therefore, it is more economical to use the two-bed system in this application, with the weak base and strong base anion exchangers in the same unit. This will yield a reduction in capital costs for the plant design.

Case D: Two-Bed Demineralization with Decarbonation, with Weak Acid and Strong Acid Cation Resins in One Stratified Bed Unit (Figure 7-4).

Design Data.   The total flow is 300 gal/min and 260,000 gal/hr. The final treated water quality is:

$$TDS \; < \; 16 \; mg/L$$

$$Conductivity \; < \; 80 \; microohms$$

$$Silica \; < \; 0.4 \; mg/L \; as \; SiO$$

The raw water is from a well, requiring no pretreatment. This two-bed system is to be an extension of a similar demineralizer. However, the existing plant uses strong acid cation exchanger alone, and there is a large excess of acid in the regenerant discharge water. The net excess acid naturally requires caustic soda for neutralization before discharge. This new demineralizer must be designed so that it will provide a net excess of caustic soda for neutralizing the excess acid from the strong acid cation exchange unit.

Design Calculations.   (a) Determination of Demineralizer System. A two-bed system with a strong base anion resin will be sufficient to supply the treated water quality as specified. To reduce the acid regenerant level to a minimum, the cation unit should have a layer of weak-acid cation resin above a layer of a strong-acid cation exchanger in a stratified unit. The decarbonator is justified because the free carbon dioxide in the cation effluent is 88% of the 242 mg/L, or 213 mg/L plus 3 mg/L present in the raw water supply, for a total of 216

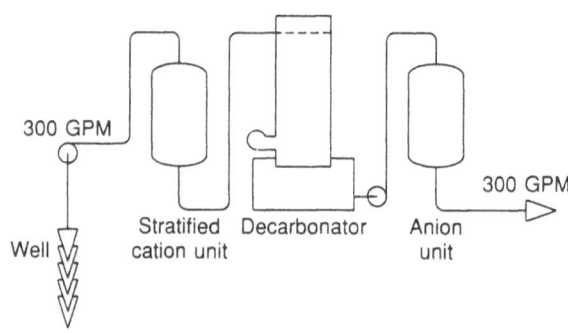

Figure 7-4.   Case D: Final process schematic.

mg/L. The decarbonator should be designed to reduce this 216 mg/L to about 5 mg/L in order to save on the caustic soda regenerant for the anion resin.

(b) Selection of Resins, Capacities and Regenerant Level: Weak Acid Cation Exchanger (upper layer). To remove the cation associated with the bicarbonate alkalinity (242 mg/L), a carboxylic cation exchange product is required. At a hardness-to-alkalinity ratio of about 0.25 and a total mineral acidity of 71 mg/L, the rated capacity is 14 to 15 kg/ft^3. However, as the capacity data indicate, when a weak acid cation exchanger is used in a stratified bed application, the capacity rating can be substantially increased. In this application, we can assume a capacity rating of 19 kg/ft^3.

Strong Acid Cation Exchanger (lower layer). To remove those cation associated with the total mineral acidity, we will use a strong acid cation exchange product. This material has a higher density than the weak acid cation exchanger and therefore will remain on the bottom of the unit during backwashing.

The specified cation leakage is 16 mg/L. Since most of the service run will result in low alkalinity and nearly 100% sodium ion in the effluent of the weak acid cation exchanger layer, an acid level somewhat over 7.5 lb/ft^3 of resin bed would satisfy the 16 mg/L requirement. To provide a safety factor we will use 10 lb/ft^3 and use a rated capacity of 18 kg/ft^3, which is 90% of the capacity rated for most commercial products.

Strong Base Anion Exchanger. The total exchangeable anions in mg/L as $CaCO_3$ are:

Free carbon dioxide: 5 mg/L × 1.13 =   6

Silica   18 mg/L × 0.83 = 15

Total mineral acidity   = 71

Total exchangeable anions   = 92 mg/L

                                      = 5.4 g/gal

$$\% \text{ Weak acids} = \frac{21 \text{ mg/L} \times 100}{92 \text{ mg/L}} = 23\%$$

$$\% \text{ Silica} = \frac{15 \text{ mg/L} \times 100}{92 \text{ mg/L}} = 16\%$$

The silica leakage of 0.4 mg/L required can be obtained with 5 lb of NaOH/ft^3 with a type-I strong base anion exchange product. At this regenerant dosage, the rated capacity is 11.7 kg/ft^3.

(c) Design of Demineralized System.

Anion Unit:

Capacity: 260,000 gal × 5.4 g/gal/1000 = 1410 kg required

$$\text{Bed volume} = \frac{1410 \text{ kg}}{11.7 \text{ kg/ft}^3} = 121 \text{ ft}^3$$

At 300 gal/min and a flowrate of 8 gal/min/ft², the required bed area is:

$$\frac{300 \text{ gal/min}}{8 \text{ gal/min/ft}^2} = 37.4 \text{ ft}^2$$

hence, we will use a 7 ft diameter shell. This will yield a bed depth of 3.2 ft.

Bed depth	=	3.2 ft
Freeboard 100%	=	3.2 ft
Straight shell	=	6.4 ft

Therefore, the anion unit will be 7 ft in diameter with a straight side shell height of 7 ft. The NaOH required for regeneration will be 5 × 121 = 605 lb as 100% NaOH.

The rinse water to waste minus the recycle will be 50 × 121 = 6050 gal.

Decarbonator: To reduce the free carbon dioxide from 216 mg/L to 5 mg/L, we will use a wooden tower 4 ft in diameter and 11 ft high, with a 5 ft layer of Rashig rings and a blower delivering 1000 standard ft³ of ambient air per min at a pressure of about 4 psi.

Cation unit (weak acid cation layer)

Anion outlet volume	=	260,000 gal
Anion waste water	=	6050 gal
1/2 anion recycle water	=	3025 gal
	Total =	269,075 gal

Alkalinity = 14.1 g/gal

Capacity = 14.1 g/gal × 269,075 gal/1000 = 3800 kg

$$\text{Weak acid resin bed volume} = \frac{3800 \text{ kg}}{19 \text{ kg/ft}^3} = 200 \text{ ft}^3$$

$$\text{Acid requirement} = 1.2 \times \frac{3800 \text{ kg}}{7} = 650 \text{ lb } 100\% \text{ H}_2\text{SO}$$

Cation unit (strong acid cation layer)

$$\text{Total mineral acidity} = \frac{71 \text{ mg/L}}{17.1} = 4.2 \text{ g/gal}$$

$$\text{Strong acid resin bed volume} = \frac{1130 \text{ kg}}{18 \text{ kg/ft}^3} = 63 \text{ ft}^3$$

This would be too shallow. Therefore, we would use a minimum of 18 in. for the depth of the resin layer; which is 75 ft³ in an 8-ft diameter vessel. This diameter is chosen to avoid too deep a resin bed.

The straight side shell height can now be estimated as follows:

Strong acid resin: 75 ft^3, 75 ft^3/50 ft^2 = 1.5 ft

Weak acid resin: 200 ft^3, 200 ft^3/50 ft^2 = 4.0 ft

20% swelling	0.8 ft
75% freeboard	4.8 ft
Anthracite subfill	1.4 ft

Straight side height = 12.5 ft

We will use one cation unit measuring 8 ft in diameter by 13 ft high. The anthracite subfill is necessary to insure even distribution of the upflow acid regenerant.

The total acid required for both cation exchangers is 844 lb. This is more than 11 lb/ft^3 on the strong acid cation exchanger and, therefore, the leakage will be minimal (well below the specified 16 mg/L).

(d) Regeneration Schedules. The acid and caustic soda regenerations take place at the same time. In this system, the acid can be injected at a concentration of 3% since there is so little calcium in the water supply. The caustic soda is injected at a concentration of 4% and a temperature of 120°F. The total regeneration period for the two regenerations is less than 3 hr.

(e) Excess Acid and Excess Caustic Soda in the Waste Regenerant:

Total acid	= 844 lb
Consumption by weak cation resin	= −545 lb
Consumption by strong cation resin	= −162 lb
Excess acid as $CaCO_3$	= 137 lb
Total caustic soda as $CaCO_3$	= 758 lb
Consumption by anion resin	= 201 lb
Residual caustic soda	= 557 lb

Therefore, the net excess caustic soda is 557 − 137 = 420 lb. The double layer weak acid/strong acid combination cation exchange units use about 1000 lb less per day than a single layer strong acid cation exchange unit. This savings amounts to $10 a day when the cost of sulfuric acid is 1 cent per pound.

Part 2

LABORATORY UNITS

UNIT 1 PART A: SHAPE, FORM, AND QUALITY

Objective

To determine the qualitative differences associated with new and used ion exchange materials.

Discussion

Microscopic examination of an ion exchange material yields qualitative results which are helpful in defining as-received quality, relative particle size distribution, and the effects of service life. While the results of this test is somewhat subjective, the data will indicate differences between manufacturers, and the structure of ion exchange materials.

Procedure

Obtain four samples of moist ion exchange resin from the laboratory instructor. Transfer a small portion of each sample to the cavity of a culture slide, wet with 1 or 2 drops of deionized water, and label each slide. Using the high-powered objective ($100\times$), visually examine each material with both top and bottom illumination. Take notes on the shape and surface appearance of each sample. Also note whether the samples are translucent or opaque in character. Using a reticule eye piece and the low-power objective ($25\times$) and a millimeter scale, estimate the particle size range and the predominant particle size. Record these particle size estimates for each material.

Finally, record the number of whole clear, whole cracked, and broken particles in the microscope field. Repeat the size measurements and quality determinations on a second or third sample taken from the same stock container.

Calculate the relative bead quality as follows:
Let

T = total number of particles counted
N_w = number of whole clear beads counted
N_c = number of cracked beads counted
N_b = number of broken pieces counted

Then

$$\% \text{ whole beads} = (N_w/T) \times 100$$

$$\% \text{ cracked beads} = (N_c/T) \times 100$$

$$\% \text{ broken beads} = (N_b/T) \times 100$$

PART B: SPHERICITY

Discussion

Microscopic examination alone will not identify split beads, since half beads on a slide will appear as whole beads in the field of view. The following technique can be used to determine the relative percent of spherical beads in a given sample of ion exchange material.

Procedure

Air-dry about 50 g of the as-received moist ion exchange sample until it is free-flowing. Obtain a 12" by 12" glass pane (with fire-polished edges) and clean it with a 1:1 solution of deionized water and isopropyl alcohol.

Raise one edge of the plate about 1" from the bench, forming a shallow inclined plane, and provide a means for catching the resin sample at the lower edge of the glass pane.

Record the weight of the sample as "A" to the nearest 0.1 g. Gently pour the air-dried resin sample near the upper edge of the glass pane, allowing the spherical beads to roll off the bottom edge. If necessary, use a camels hair brush on the concentrated spots, which may occur, freeing trapped spherical beads.

Record the weight of the spherical particles as "B" to the nearest 0.1 g, and calculate the % spherical beads as follows:

$$\% \text{ spherical beads} = (B/A) \times 100$$

Microscopic examination of the collected fraction will indicate the relative amount of cracked spherical beads.

At the conclusion of this experimental unit, your notebook must contain the following information for each sample: a qualitative statement of sample characteristics (i.e., shape, surface quality, color, opaqueness or transparency, etc.); maximum and minimum size in millimeters; estimation of predominant particle size; and duplicate or triplicate values for whole clear beads, cracked beads, and pieces. In addition, the completed results of the sphericity test must be turned in with answered questions for grading.

QUESTIONS

1-1. Are the observed variations sufficient to identify different products? Give at least two examples.

1-2. Calculate the relative precision of the duplicate bead quality measurements.

1-3. Could this precision be improved? Give an example.

1-4. As a qualitative measurement, do we need to improve this precision? Discuss.

1-5. Which samples were found to be translucent? and opaque?

1-6. Explain in general terms why some beads stick to glass or plastic surfaces?

1-7. Do the observations tell us anything about the ionic form of the materials; or define anion exchange products from cation exchange products? Explain.

1-8. Would visual observations of this kind be helpful to define the relative effects of:

    (a) Wetting–drying cycles?

    (b) Freeze–thawing cycles?

    (c) Operating cycles?

    (d) Long–term stability?

1-9. Considering your answers to question 1-8, outline a test procedure to define the effects of (a), (b), (c), and (d).

1-10. Assuming that a specific application will degrade an ion exchanger, give two simple measurements that might be used to define the degree of degradation as function of operating time?

UNIT 2   DENSITY

Objective

To determine the relationship among as-shipped bulk density, minimum tapped-down density, and absolute density of a given ion exchange material.

Discussion

Manufacturers of ion exchange materials and contributors to the technology of ion exchange have used several different methods for the determination of product density. At the present time, manufacturers use wet-bulk density (lb/ft^3 or kg/m^3) as a basis for shipping their products. Laboratory evaluations are generally expressed in g/ml, where the volume (ml) in this case is the minimum tapped-down volume of a wet sample under water. The separation of two different ion exchange materials by backwashing in a mixed-bed application, however, is dependent on the absolute density of the materials with respect to water.

Therefore, the determination of density is of some importance. Also certain types of ion exchange degradation can be quantified by careful measurements of the density changes which may occur during the operating life of an ion exchange material.

PART A: APPARENT, TAPPED-DOWN, AND TRUE DENSITY

Procedure

Carefully weigh a clean, dry 50 ml graduate and record the weight as "A" to the nearest 0.1 gram. From the as-received moist supply of ion exchange material available in the laboratory, transfer about 20 g to the preweighted graduated cylinder, reweigh, and record the second weight as "B" to the nearest 0.1 g.

Add sufficient deionized water to flood the sample and stir with a thin glass rod to dislodge entrapped air bubbles. Add enough deionized water to fill the graduate to the 50 ml level, reweigh, and record the weight as "C" to the nearest 0.1 g.

Mix the sample with a thin glass rod and let it settle undisturbed for 5 min. Record the settled volume as "D" to the nearest whole milliliter. Using a rubber stopper, tap the sides of the graduate while noting the volume of the particulate solids. Record the minimum tapped-down volume as "E" to the nearest whole milliliter.

Clean the 50 ml graduate, dry the outside, and fill to the 50 ml level with deionized water, being sure to remove all air bubbles. Record the weight of the graduate with the deionized water as "F" to the nearest 0.1 g.

Calculations

You now have the data required to calculate the apparent density, tapped-down density, and the true density of the ion exchange material as follows:

$$\text{Apparent density, wet g/ml} = (B - A)/D$$

$$\text{Tapped-down density, wet g/ml} = (C - A)/D$$

$$\text{True density, wet g/ml} = (B - A)/[F - A - (C + B)]$$

PART B: BULK DENSITY OR SHIPPING WEIGHT

Procedure

Set up the apparatus as shown in Fig. U2-1. Fill the addition funnel with deionized water, which has been boiled and cooled to reduce the presence of dissolved gases. Adjust the stopcocks at top and bottom so that water flows up through the empty column and out the backwash line. Allow all the water to flow through to displace all the air. It may be necessary to tap the side of the column to dislodge any air bubbles adhering to the walls. Stop the flow when the water is near the bottom of the addition funnel, and remove the stopper and stopcock from the top of the column.

Carefully drain the water from the column until the level reaches the coarse frit support plate. Add deionized water in 10 ml portions using a class A 10 ml pipette, and measure the distance from the top of the frit to the water miniscus after each addition using a centimeter scale. Record the measurements to the nearest 0.1 cm as a function of the volume added for the calibration of the column.

Drain the water half way down the column. Weigh out about 30 g of the as-received moist ion exchange material supplied by the laboratory instructor and record the weight as "A" to the nearest 0.1 g. Transfer this sample to the column and close the top with the stopper removed earlier.

Refill the addition flask with boiled and cooled deionized water and start backwashing. Increase the flowrate until the fluidized bed volume is about twice the initial settled volume. As quickly as possible and in the order shown:

—Shut off flow from the addition flask
—Open top of column
—Open bottom drain

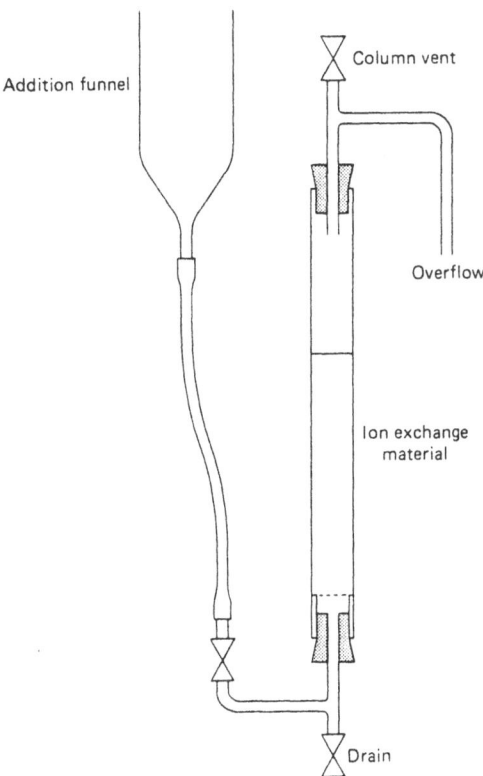

Figure U2-1. Part B: Bulk density.

Allow the resin bed to drain down to a fixed level without jarring and shut off the drain flow. Record the distance from the bottom frit to the top of the resin bed as "B" to the nearest 0.1 cm. Repeat the measurement of backwashed and drained volume at least three times. Using the calibration data, convert the depth measurements into bed volume expressed as milliliters.

Calculations

For each measurement do the following calculation:

$$\text{Bulk density,} \quad \text{lb}/\text{ft}^3 = (A/B) \times 62.4$$

Also, calculate the mean and standard deviation for the measurements made.

Starting with part A, repeat the experiment with a different ion exchange material.

Your notebook must contain the following information:

—Sample identification
—All weights to the nearest 0.1 g
—Calibration data
—Final calculations

Appropriate data and answered questions will be submitted to the laboratory instructor for approval and grading.

QUESTIONS

2-1. The difference between the apparent density and the tapped-down density is related to the degree of packing. Explain.

2-2. Using Stoke's Law,

$$V = \text{ms}/\text{sec} = \frac{2gr^2(dr - dw)}{9\mu}$$

where, without temperature corrections, we have:

g = acceleration of gravity, 980 cm/sec
μ = the viscosity of water, 0.01002 g/cm/sec
r = partical radius, 0.02 cm
dw = density of water, 1.000 g/cm^3
dr = density of the ion exchanger
    (a) Anion resin = 1.09 g/cm^3
    (b) Cation resin = 1.30 g/cm^3

Calculate the settling rate of the anion and cation exchanger materials.

2-3. Can mixtures of anion and cation resins in question 2-2 above be separated by backwashing in a column?

2-4. How might the measurement of backwashed and drained (B&D) volume be improved?

2-5. With the B&D volume known, what is the volume of 2% sulfuric needed to regenerate an experimental bed volume of 300 ml at a dosage 10 lb/ft³ (assume 100% sulfuric acid)?

2-6. With a shipping weight of 50 lb/ft³ (802 kg/m³), how many ft³ of material must be purchased, to the nearest whole ft³, for a commercial unit 3 ft inside diameter and a bed depth of 30 in.? (Allow 80% for freeboard.)

2-7. At $76.00 per cubic foot as shipped, what is the cost of the ion exchange material required in question 2-6?

2-8. Removal of iron impurities from concentrated $AlCl_3$ solutions (specific gravity $\geq$ 1.18) is a commercial process. How will this effect operation? Explain.

2-9. A condensate polisher measuring 8 ft in diameter has a bed depth of 36 in. What is the weight of resin in the vessel (see question 2-2 and assume an anion/cation ratio of 1.6)?

2-10. In question 2-9 above, freeboard is 20% and the vessel weight is 4000 lb. With 40% void volume, estimate the load on the four support legs during operation (water density = 62.4 lb/ft³).

UNIT 3   VOLUME CHANGES

Objective

To evaluate the relative effect of volume changes of ion exchange materials.

Discussion

Ion exchange materials are, in a sense, solid solutions with polar groups on a three-dimensional matrix. All ion exchangers hold enough water to hydrate the ion held by the polar group.

This degree of hydration is dependent on:

—The counter ion attached to the polar group
—The cross-linking, or degree of rigidity, of the polymer
—The concentration of the electrolyte in the pore volume of the ion exchange material

Therefore, ion exchange materials undergo volume changes during normal service operations. This experimental unit is designed to demonstrate some of the effects on the bulk volume properties of ion exchange materials.

PART A: EFFECT OF CONCENTRATION

Procedure

Obtain a sample of low cross-linked cation exchange material from the labo-ratory instructor. Weigh out four 30.0 ± 0.1 g portions. Transfer one sample as received to the column used in Unit 2. Transfer the remaining samples in-dividually to separate 100 ml beakers, and equilibrate each individually in 5% NaCl, 10% NaCl, and 15% NaCl by soaking each at least 1 hr at room tem-perature.

Carefully determine the B&D bed volume of the as-received sample using the technique learned from Unit 2. Record the average bed volume as ''A'' to the nearest 0.5 milliliter. Remove the ion exchange material from the column and clean the column. Collect all the used ion exchange material from this experiment in a 250 ml beaker and return it to the laboratory instructor.

Quantitatively transfer the sample and 5% NaCl solution to the empty column. Fill the addition funnel with 5% NaCl solution. Determine the B&D bed volume using the 5% NaCl solution. Record the average bed volume as ''B'' to the nearest 0.5 ml.

Repeat the B&D measurement on the sample stored in 10% NaCl back-washing as before with 10% NaCl from the addition funnel. Record the average bed volume as ''C'' to the nearest 0.5 ml.

Finally, determine the B&D bed volume of the sample stored in the 15% NaCl solution. Record the average bed volume observed as ''D'' to the nearest 0.5 ml.

Calculations

Calculate the volume change for each concentration of NaCl used from the averaged data as follows:

For 5% NaCl, $(A - B) \times 100/A =$ _____ %

For 10% NaCl, $(A - C) \times 100/A =$ _____ %

For 15% NaCl, $(A - D) \times 100/A =$ _____ %

Plot the calculated % change as a function of the NaCl concentration.

PART B: EFFECT OF IONIC FORM

Procedure

Obtain a sample of a weak acid cation exchange material from the laboratory instructor. Weigh out about 20 g and quantitatively transfer the weighed sample to the previously calibrated column used in Unit 2. Using deionized water, determine the as-received B&D volume, and record the average bed volume as

"A" to the nearest 0.5 ml. Drain the excess deionized water from the addition funnel.

Prepare a dilute solution of sodium hydroxide by dissolving 4.00 g of NaOH in 4 L of deionized water (this equal to $1000/40.08 = 2.495$ meq Na^+ ion/L). Transfer 1 L of the dilute NaOH solution to the addition funnel and backwash the sample at a flowrate of about 20 ml/min. Follow the NaOH solution with a 300 ml backwash rinse of deionized water. Determine the new B&D bed volume using deionized water, and record the average bed volume as "B" to the nearest 0.5 ml.

Repeat this procedure four more times, and record the B&D bed volume after each treatment as "C," "D," "E," and "F," respectively.

Calculations

Calculate the volumetric change as a function of the milliequivalents of sodium ion passed through the sample bed for each treatment as follows:

at   2.495 meq Na^+ $(B - A) \times 100/A =$ _____ %

at   4.990 meq Na^+ $(C - A) \times 100/A =$ _____ %

at   7.485 meq Na^+ $(D - A) \times 100/A =$ _____ %

at   9.980 meq Na^+ $(E - A) \times 100/A =$ _____ %

at   12.475 meq Na^+ $(F - A) \times 100/A =$ _____ %

Plot the percent volume change as a function of the number of meq of Na^+ ion passed through the ion exchange material in the column.

PART C: EFFECT OF HYDRATION

Procedure

Obtain a sample of a gel type and a macroporous type strong acid cation exchange material from the laboratory instructor. Determine the relative quality of each sample while still moist, using the techniques learned in Unit 1. Record the % whole, % cracked, and % broken beads for the as-received condition of each material.

Dry each material overnight at 105°C. Remove the samples from the oven, and allow them to cool to room temperature. Transfer a few grams of each material to 200 ml beakers. Add 100 ml of deionized water to one and label. Add 100 ml of 15% NaCl to the second portion and label. Stir each sample by swirling (do not use a glass rod or stirring device), and let stand for 1 hr. Pour off the excess solution and redetermine the bead quality, recording the % whole, % cracked, and % broken found for each material and each solution used. All observation and calculations must be turned in to the laboratory instructor with the answered questions for approval and grading.

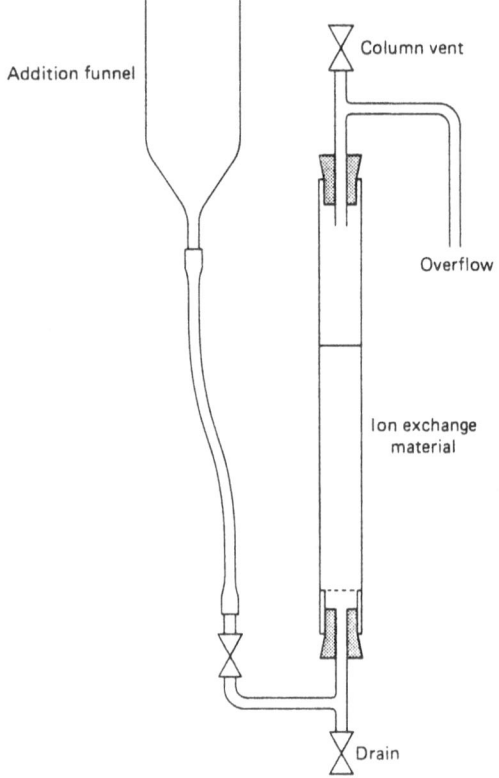

Figure U3-1.   Part A: Effect of concentration. Part B: Effect of ionic form.

QUESTIONS

3-1. Estimate the volume change in a 65 ft^3 (1.84 m^3) softener after regeneration with 6 lb NaCl/ft^3 (96.3 kg/m^3) at a concentration of 12%.

3-2. Could the volume change be used to estimate the degree of ion exchanger exhaustion? Explain. (Note: Resin capacity is 3.5 meq/ml of B&D bed volume.)

3-3. When the ion exchanger is rinsed to a residual of 2% regenerant as indicated in question 3-1 above, will there be a volume change? If so how much?

3-4. Could this experiment be done downflow in a glass column? If not, why not?

3-5. Can the volume change during regeneration be used to estimate resin degradation as a function of operating life? Explain why.

3-6. A unit containing 5 ft^3 of R—COOH resin is used to neutralize a caustic waste of unknown concentration. After processing 600 gal-

Ions upflow, the volume change is found to be 60%. What is the normality of the caustic waste solution? Show all calculations.

3-7. Assuming a capacity of 10.4 meq Na ion/ml of bed (1 ft³ = 28,300 ml), is the resin bed exhausted in question 3-6 above?

3-8. Estimate the remaining volume of caustic waste solution that can be processed by the bed in question 3-6 above and show all your calculations.

3-9. An R—SO₃H resin has a volumetric capacity of 1.95 meq/ml. What is the estimated treatment volume, expressed in gallons, given the answer from question 3-6?

3-10. Could a strong acid cation exchanger in the hydrogen form be used to neutralized the caustic waste defined in question 3-6? Explain your answer.

UNIT 4   POROSITY OF ION EXCHANGE MATERIALS

Objective

To demonstrate the differences between gel structure and macroporous structure ion exchange materials.

Discussion

When ion exchange materials are used to treat surface water supplies, they may be exposed to a wide variety of soluble and colloidal organic compounds. In most cases, these organics have high molecular weights, and are large polyelectrolytes. The relative porosity of the ion exchange material is an aid to applications of this kind. While gel products are clear and transparent, and porous products in general are opaque, it is not always possible to determine the difference visually.

The size of the pores affect the capacity, quality of effluent, and volume of rinse, since ion exchange materials with small pores have been known to adsorb polar organic species irreversibly leading to low operating capacities and long rinses. This experimental evaluation of pore size will give relative indications of the degree of porosity. The relationship between pore volume, V, pore radius, r, and specific surface area, A, is:

$$r_{cm} = \frac{V_{cm}{}^3}{A_{cm}{}^2}$$

There are instruments available commercially which are capable of measuring the parameters in the above equation. In general, they are expensive and unavailable in most laboratories. Some of the techniques used are:

- Surface area BET method (Brunauer–Emmet–Teller)
- Pore volume saturation absorption with organic solvents
- Size distribution high-pressure mercury intrusion

In addition, information on pore size and distribution can be obtained by using an electron microscope, where position, appearance, and character of the pores can be estimated and described.

Procedure

Obtain a sample of a typical gel type and a typical macroporous ion exchange material from the laboratory instructor. Wash each several times with deionized water, and remove the excess water from each by aspirator suction. Using previously tared filter crucibles, transfer each sample to a drying oven and dry overnight at 105°C.

Remove the samples from the oven and cool them to room temperature in a desicator. Weigh and record the net dry sample weight of each as "A" to the nearest 0.001 g.

Place the crucibles with the dried resin samples in separate 200 ml beakers in a desicator. Add enough n-Heptane, $CH_2(CH_2)_3CH_2$, to cover the ion exchanger samples in the crucibles. Close the desicator and remove any entrained air by connecting the desicator to a water spirator for 2 min. Shut off the aspirator and slowly allow the air back into the desicator. Open the desicator, remove each sample, and place each on an operating aspirator for exactly 10 sec.

Immediately determine the total weight and record the net sample weight with the adsorbed heptane as "B" to the nearest 0.001 g.

The best determination is obtained by taking a second tare after soaking the empty crucible in Heptane for 1 min followed by the 10 sec aspiration process.

Calculation

Calculate the pore volume of each ion exchange material as follows:

$$\text{Pore volume,} \quad \text{ml/dry g} = (B - A)/A \times p$$

where p = density of heptane at 25°C (0.703 g/ml) *Note*: The high molecular weight organic anions found in water not only diffuse out very slowly, but also due to their aromatic nature become "adsorption-bonded" to the aromatic groups on the ion exchangers. This bond is not broken during regeneration.

At the completion of this experimental unit you must turn in all data, including the weights, calculations, and answered questions to the laboratory instructor for approval and grading.

QUESTIONS

4-1. Are ion exchange groups on the surface of the bead or distributed throughout the bead material?

4-2. In order to react, is contact between the exchange site and the polar group necessary?

4-3. Which will have a higher diffusion rate, a small ion or one with a high molecular weight?

4-4. Can fouling be explained on diffusion rates only, or are other factors involved? Explain.

4-5. Give at least three other examples of nonpolar organic compounds that could be used to evaluate the porosity of ion exchangers.

4-6. Does the test procedure tell us anything about the pore size distribution?

4-7. What equipment is required to define the pore size distribution and absolute pore volume?

4-8. Given a choice, what ion exchange material would you suggest for treating a surface water supply without pretreatment? Explain why.

4-9. With pretreatment, what would you suggest for question 4-8 above? Explain why.

4-10. What two operating criteria can be used to define the effects of fouling?

UNIT 5 HYDRAULIC PROPERTIES

Objective

To demonstrate the effect of particle size on the backwash expansion and pressure drop of a confined bed of ion exchange material.

Discussion

The hydraulic properties of an ion exchange material are important physical parameters, since they are used to determine the operating unit height and pumping requirements. As a result, they bear directly on the capital cost for a given ion exchange application.

These properties are usually available from ion exchange manufacturers and are frequently encountered in ion exchange applications literature. The backwash expansion and pressure drop are related to the following properties:

- Particle size
- Flowrate of process stream

- Density of ion exchanger
- Temperature of process stream
- Viscosity of process stream
- Density of process stream

In general, the data are presented in the literature in two different forms.

Bed Expansion Versus Flowrate

The degree of expansion, expressed as the percent of bed height as a function of the backwash flowrate, is used to determine the "freeboard" required. The depth of the ion exchange bed plus the required "freeboard" defines the straight-side height of the ion exchange unit.

Pressure Drop Versus Flowrate

This information is used to determine if auxilliary pumps are needed in an ion exchange application. In addition, any increase in pressure drop during the service life of an ion exchange material may indicate some physical breakdown of the ion exchange material, or the accumulation of solid debris from the effluent.

PART A: BED EXPANSION VERSUS FLOWRATE

Procedure

Record the laboratory temperature. From a sample of moist ion exchange material designated by the laboratory instructor (screened to pass a 16 mesh screen and be retained on a 20 mesh screen), measure out about 40 ml and transfer the measured portion to the column shown in Figure U5-1. Add deionized water to the column from the addition funnel before adding the ion exchange material to the column. Check Figure U5-1 for the proper valve settings.

Allow the resin to settle without jarring and record the bed depth at rest, H_0, to the nearest 0.1 cm. Fill the addition funnel with deionized water at room temperature. Slowly let water into the column from the addition funnel to displace all the air through the top valve and close this vent.

Open the overflow valve and start backwash flow at a low flowrate. Determine the flowrate by the amount of water collected in a suitable graduated cylinder for 1 min. Without changing the backwash flowrate, measure the depth of the expanded bed and record the nearest 0.1 cm with the flowrate measured at the same time.

Increase the backwash flowrate increments and record the bed depth and flowrate after each change. Carry out enough measurements while increasing or decreasing the backwash flowrate to establish a curve relating the backwash flowrate to the observed bed expansion.

After the last observation, allow the ion exchange material to settle undisturbed and recheck the value of H_0, to the nearest 0.1 cm, to assure that no material was lost from the column.

Calculation

At each backwash flowrate, calculate the percent expansion of the ion exchange bed and arrange the results in a table which shows the relationship between flow rate and bed expansion. The equation for this calculation is:

$$\text{Bed expansion, } \% = (H - H_0) \times 100/H_0$$

where H is the individual bed depth measurements made at several different backwash flowrates.

PART B: PRESSURE DROP VERSUS FLOWRATE

Procedure

With the ion exchange material still in the column, arrange the valve settings and connections as shwon in Fig. U5-2. Verify that the manometer legs are at the same height. Measure the distance between the center lines of the side taps and record the measurement as d, to the nearest cm.

Before starting, be sure all air has been removed from the manometer legs, and that the addition funnel is filled with deionized water at the prevailing room temperature.

Start the water downflow through the ion exchange material, and increase the flowrate until some change is observed in the manometer. Measure the flowrate by using a suitable graduate and by collecting the effluent for exactly 1 min. Without changing the flowrate, measure and record the displacement in the manometer to the nearest 0.1 cm.

Increase the flowrate in small increments and record the flowrate and manometer displacement after each change. Continue the measurements until the relationship between flowrate and pressure drop has been established to your satisfaction.

Calculation

At each flowrate, calculate the pressure drop and record the data in a table which shows the relationship between the observations. The following equations can be used to calculate the pressure drop at each flowrate:

$$\text{Pressure drop, } \text{psi/ft of bed} = (h \times 5.119)/d$$

where h = the manometer reading to the nearest 0.1 cm, and d = the distance between the side taps to the nearest 0.1 cm.

The constant 5.119 assumes that the manometer is filled with mercury. When a dense oil is used, this constant can be adjusted for the density difference as follows:

$$5.119 \times \frac{\text{Density of mercury}}{\text{Density of oil used}}$$

Repeat the test procedure with the same ion exchange material, which has been prescreened to pass a 30-mesh screen and be retained on a 40-mesh screen. Prepare a graph showing the relationship between backwash flowrate and percent bed expansion for both the coarse and fine ion exchange materials. Also, prepare a similar graph for the pressure drop and flowrate observations.

Tabulated data and graphs must be turned in with the answered questions to the laboratory instructor for grading.

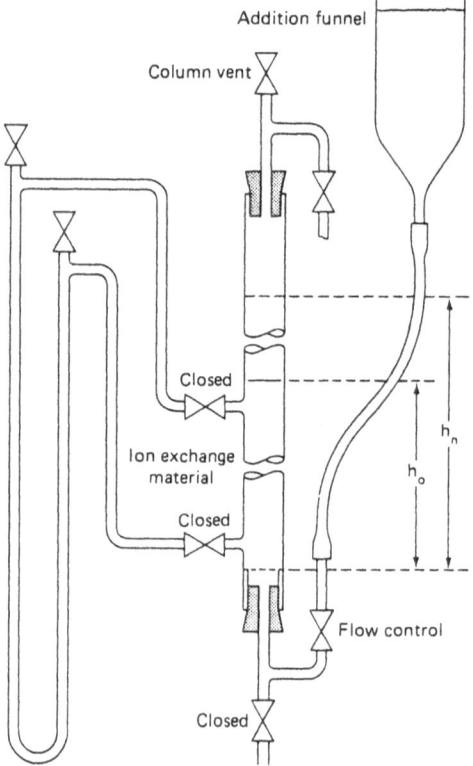

Figure U5-1. Hydraulic properties. Part A: Bed expansion.

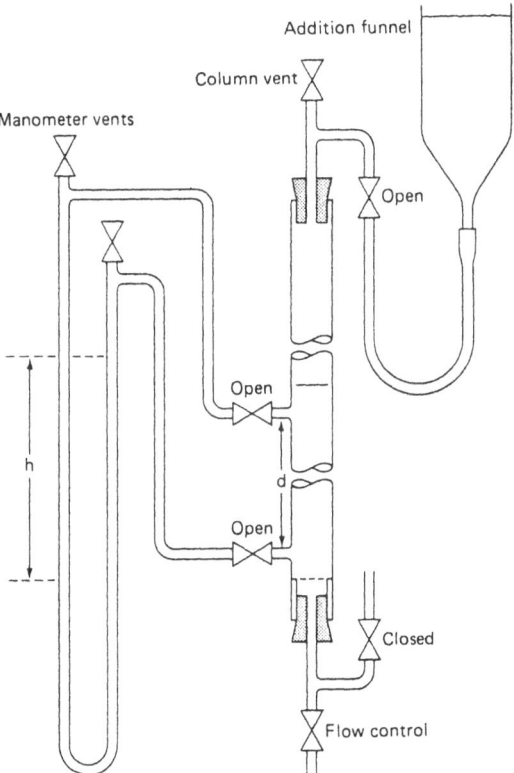

Figure U5-2. Hydraulic properties Part B: Pressure drop.

QUESTIONS

5-1. Why is it essential to know the backwash rate for various sizes of ion exchange particles?

5-2. Which resin particles would be lost if the backwash flowrate was too high?

5-3. Which particles would remain on the bottom if the backwash flowrate was too low?

5-4. What is the indication if the resin bed volume is reduced after backwashing? What if the bed volume increases after backwashing?

5-5. Give at least two reasons for backwashing an ion exchange bed.

5-6. If pressure drop increases during service life, what can be the reason for the increase?

5-7. Will the formation of fines increase or decrease the pressure drop?

5-8. Can pressure drop be due to other causes than the formation of fines? Give at least three reasons for increases in pressure drop exclusive of the formation of fines.

5-9. Given a pressure drop of 0.75 psi/ft of bed depth, at a flowrate of 8 gal/ft^2, what is the pressure drop on a unit containing 150 ft^3 with an inside diameter of 7 ft?

5-10. Given a bed expansion of 40% at a backwash rate of 6 gal/min/ ft^2, what is the minimum free board requirement to prevent resin loss?

UNIT 6   IDENTIFICATION OF FUNCTIONAL GROUPS

Objective

To qualitatively identify the function group on an unknown ion exchange material before the determination of total capacity or other chemical characteristics are evaluated.

Discussion

With experience, an initial identification of the general ion exchanger type can sometimes be made by microscopic examination; however, visual examination is not always certain. When an ion exchange material is unknown, auxiliary qualitative tests may be required in order to identify the functional group (i.e., strong, weak, anion, or cation). Functional group identification is necessary, since data to be obtained later is dependent on the reference form of the ion exchanger under examination.

This qualitative test is based on the following equations:

- $R-SO_3H + NaNO_3 = R-SO_3Na + HNO$
- $R-COOH + NaNO_3 = R-COOH + NaNo_3 + trace\ H\ ion$
- $R-N(CH_3)_3Cl + NaNO_3 = R-N(CH_3)NO_3 + NaCl$
- $R-N(CH_3)_2-HCl + NaNO_3 = R-N(CH_2)_2-HCl + NaNO_3$
                                               $+ trace\ Cl$

where $R-SO_3H$ is the stron acid cation exchanger; $R-COOH$ is the weak acid cation exchanger; $R-N(CH_3)_3Cl$ is the strong base anion exchanger; and $R-N(CH_3)_3-HCl$ is the weak base anion exchanger.

Procedure

Transfer a few grams of each of the unknown resins assigned by the laboratory instructor to 100 ml beakers. Add 50 ml of 10% HCl to each; stir, cover, and allow to stand at room temperature overnight.

Decant and discard the excess HCl solution. Rinse each sample by repeated additions of deionized water until pH of the rinse water in contact with the sample is less than 4.5 (methyl orange can be used to indicate the completeness of rinse). To each unknown, add 50 ml of 10% $NaNO_3$ solution; stir, and allow 1 hr for equilibration.

Determine and record the pH of the $NaNO_3$ in contact with each sample. Transfer a small amount of the supernate from each unknown to a clean test tube, add 1-2 drops of 1% $AgNO_3$, and record the relative amount of chloride precipitate (i.e., trace, slight, or copious, etc.).

Use the following table to identify the functional group of the unknown sample:

Ion Exchanger Type	Functional Group	Response	
		pH	AgCl Precip.
Strong cation	$R-SO_3$	Low <2	None
Weak cation	$R-COOH$	Low <4	None
Strong anion	$R-N(CH_3)_3$	About neutral	Copious
Weak anion	$R-N(CH_3)_2$	Low <4	Very slight

Calculations

Since this is an experiment which yields only qualitative information, there are no calculations.

Your notebook should contain information related to all the observations made. The following table should be filled out for those coded samples assigned by the laboratory instructor. When completed, the table below and the answered questions should be turned in for grading.

Sample No.	Code No.	Functional Group			
		$R-SO_3^{-2}$	$R-COOH$	$R-N(CH_3)_3^+$	$R-N(CH_2)_2$
6-1	-----------	------------	-------------	---------------	--------------
6-2	-----------	------------	-------------	---------------	--------------
6-3	-----------	------------	-------------	---------------	--------------
6-4	-----------	------------	-------------	---------------	--------------

QUESTIONS

Complete and balance each of the following equations:

6-1. $R-SO_3H^+ + NaCl =$ _____ + _____ .

6-2. $R-N(CH_2)Cl^- + Na_2SO_4 = $ _____ + _____ .

6-3. $R-N(CH_2)-HCl + NaNO_3 = $ _____ + _____ .

6-4. $R-COOH + NaOH = $ _____ + _____ .

6-5. $R-COOH + HNO = $ _____ + _____ .

6-6. An unknown sample yields no precipitation with $AgNO_3$, and the pH of the $NaNO_3$ solution on the resin is 4.3. What is the probable functional group? What reference form should be used for additional analytical work?

6-7. Could this technique be used to identify the components in a mixed bed without separation? If not, why not? Show the appropriate equations.

6-8. After separation of a conventional mixed bed, outline a test that will identify each component.

6-9. An unknown sample from the field yields a copious precipitate with dilute silver nitrate when this test is used. What is the most probable identification?

6-10. Two samples yield faint precipitates with silver nitrate solutions. What additional piece of data is required to identify the ion exchanger types?

UNIT 7   WATER RETENTION

Objective

To define the relative effect of structure, and functional group on the determination of water retention (sometimes called water regain).

Discussion

The water retention of an ion exchange material is a chemical property and is dependent on:

—Amount of cross-linking in the polymer matrix
—Type of functional group attached to the matrix
—The ionic form of the ion exchange material

The water retention measurement is usually made on the fully hydrated (swollen) standard ionic form of the ion exchange material. It is common practice to define the reference forms of ion exchangers as follows:

Ion exchange functional group	Reference ionic form
$R-SO_3^-$ (strong acid)	Sodium (Na)
$R-COOH$ (weak acid)	Hydrogen (H)
$R-N(CH_3)_3^+$ (strong base)	Chloride (Cl)
$R-N(CH_2)_2$ (weak base)	Free base

These particular ionic forms have been chosen because they are the ionic forms that ion exchange manufacturers use to ship their products, and these ionic forms do not degrade appreciably when dried at 100–110°C.

Water retention has become a useful diagnostic tool in the field of ion exchange technology. Increases observed in the water retention of used ion exchange materials compared to new material from the same lot might indicate structural changes and loss of cross-linking in the polymer matrix. Decreases in water retention would indicate the loss of functional groups, or fouling which would keep the functional group from taking part in the desired ion exchange reaction.

Procedure

From each of the unknowns supplied by the laboratory instructor, weigh out 2 samples of about 5 g each. Transfer each sample to a 100 ml beaker, add deionized water, and soak for about 1 h. Using deionized water, transfer the wet sample to a previously tared filter crucible and dewater with an aspirator. Continue the aspirator suction for exactly 10 min after the water has disappeared from the surface of the sample in the crucible. Record the net wet weight of each sample taken as "A" to the nearest 0.01 g. Repeat this process for each unknown sample.

After filtration, dry all samples overnight at 105°C. Cool the samples in a desicator and determine the dry weight of each as "B" to the nearest 0.01 g.

Calculation

Calculate the water retention of each sample, and the precision for each duplicate determination as follows:

$$\text{Water retention, } \% = [(A - B) \times 100]/A$$

$$\text{Precision, } \% = [(\text{value1} - \text{value2}) \times 100]/\text{average}$$

Your notebook must contain data for sample identification, all weights, including tares, and a verification of the oven temperature used for drying. Obtain the sample identifications from the laboratory instructor and report the result in tabular form. Turn in all data and answered questions for grading.

QUESTIONS

7-1. Why are standard reference forms necessary for ion exchange products?

7-2. For the same test conditions on the same product, what might an increase in water retention indicate? What would a decrease indicate?

7-3. Let C = total capacity 4.8 meq/dry g, and Hn = hydration number 11.5 mmol water/meq. Estimate the capacity loss when the measured water retention decreases from 50% to 42% using the following equation:

$$Hn = \frac{\dfrac{\%\ \text{water retention} \times 10}{18}}{N \times \left(1\ \dfrac{\%\ \text{water retention}}{100}\right)}$$

Hint: Solve for N, % loss = $[(4.8 - N)/4.8] \times 100$.

7-4. Give at least three basic material properties that may have an effect on the water retention measurement.

7-5. Loss of cross-linking due to severe operating conditions will increase or decrease the water retention? Give your reasoning.

7-6. Strong base anion exchange products will lose capacity when exposed to water temperature above 140°F. Can the determination of water retention be used to follow this trend.

7-7. In question 7-6 above, what other test or tests might be used to supplement the effect of high-temperature exposure?

7-8. Give three tests which can be used to determine the relative structural damage in an ion exchange sample. Explain each.

7-9. The word ''relative'' in question 7-8 above is important. Explain ''relative'' to what?

7-10. Hydrogen ion has a larger hydrated radius than sodium ion. Regeneration of a strong acid cation exchanger with HCl and rinsing will cause the water retention to decrease or increase?

UNIT 8   FUNCTIONALITY OF ION EXCHANGE GROUPS

Objective

To determine the functionality of ion exchange materials, and to estimate the proportion of strong, medium, and weak activities present in a given sample.

Discussion

While Unit 7 gives a technique for the identification of functional groups, it does not yield any information concerning the relative strength or distribution of the ion exchange activities on the polymer matrix.

Strong acid cation and strong base (type I or type II) anion exchange products, such as weak acid cation exchangers, are usually monofunctional. Weak base anion exchange products have been manufactured with several different and sometimes mixed diamines. As a result, these products are useful in some special applications. A knowledge of their capacity as a function of pH is important for many of these special applications.

Procedure

Samples of moist, fully regenerated unknown ion exchange materials will be assigned by your laboratory instructor. Use a small portion of the sample and the procedure in Unit 6 to identify the probable functional group or groups. The functional group(s) found will be used as a guide to the appropriate section of this procedure as follows:

Use Part A for groups found to be cationic in character.
Use Part B for groups found to be anionic in character.

PART A: CATION EXCHANGE GROUP(S)

Prepare a 0.1 normal solution of potassium chloride by weighing out 7.456 ± 0.001 g of dry potassium chloride (KCl) and transfer it quantitatively to a clean 1 L volumetric flask. Add demineralized water to dissolve the salt, fill to the calibration mark, and mix well.

Weigh out 2.00 ± 0.01 grams of the assigned sample and transfer it quantitatively to a 400 ml beaker. Add 100 ml of the 0.1 N KCl solution to the beaker containing the sample, a magnetic stirrer bar, and a pH electrode which has been connected to a pH meter.

While stirring gently, titrate the resin/water slurry by adding small portions of a standardized 0.1 N sodium hydroxide solution using a 50 ml burette (note 1 ml = 0.1 meq). After each addition, allow sufficient time for the pH to stabilize. Record the steady-state pH, and the titration volume in ml after each addition in a suitable table. Prepare a graph which relates the volume of NaOH added, in ml on the horizontal axis and the pH on the vertical axis.

Calculation

Using the graph drawn from your observations, calculate the estimated capacity at each sharp change in pH as follows:

$$\text{Capacity } \frac{\text{meq}}{\text{wet gram}} = \frac{\text{NaOH vol ml} \times \text{norm of NaOH}}{\text{sample weight, wet grams}}$$

Identify and report the pK value(s) from the graph.

PART B: ANION EXCHANGE GROUP(S)

Prepare a 0.1 normal solution of potassium chloride by weighing out 7.456 ± 0.001 grams of dry potassium chloride (KCl) and transfer it quantitatively to a clean 1 L volumetric flask. Add demineralized water to dissolve the salt, fill to the calibration mark, and mix.

Weigh out 2.00 ± 0.01 g of the assigned sample and transfer it quantitatively to a 400 ml beaker. Add 100 ml of the 0.1 N KCl solution to the beaker containing the sample, a magnetic stirrer bar, and a pH electrode which has been connected to a pH meter.

While stirring gently, titrate the resin/water slurry by adding small portions of a standardized 0.1 N sulfuric acid solution using a 50 ml burette (note 1 ml = 0.1 meq). After each addition, allow sufficient time for the pH to stabilize. Record the steady-state pH, and the titration volume in ml after each addition in a suitable table.

Prepare a graph which relates the volume of H_2SO_4 added, in ml on the horizontal axis and the pH on the vertical axis.

Calculation

Using the graph drawn from your observations, calculate the estimated capacity at each sharp change in pH as follows:

$$\text{Capacity } \frac{\text{meq}}{\text{wet gram}} = \frac{\text{NaOH vol ml} \times \text{norm of NaOH}}{\text{sample weight, wet grams}}$$

Identify and report the pK value(s) from the graph.

Hints:

- Some samples may require more time than others for the pH to become stable after each addition.
- When the pH change begins to increase or decrease after each addition, the volume of the titration should be decreased. This will help define the pH transient and provide better accuracy for estimating the capacity and pK values.

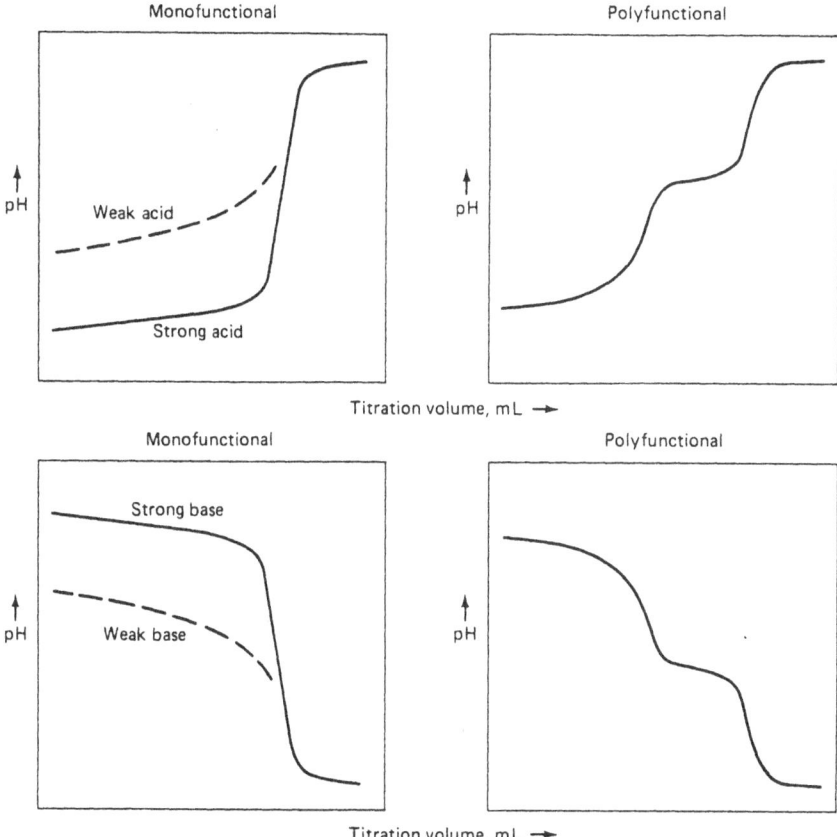

Figure U8-1. Functionality of ion exchange groups.

At the completion of this experimental unit you must turn in all data, including weights, calculations, graphs, and answered questions to the laboratory instructor for approval and grading.

QUESTIONS

8-1. Complete the following cation exchange reactions:

$$R-SO_3H + KCl + NaOH =$$
$$R-COOH + KCl + NaOH =$$

8-2. Complete the following anion exchange reactions:

$$RN-(CH_3)_3OH + KCl + HCl =$$
$$RN-(CH_3)_2 + KCl + H_2SO_4 =$$

8-3. Given the following titration data, identify the basic ion exchange material and the probable pK value(s).

Titer, ml of 0.15 N NaOH	pH	Titer, ml of 0.15 N NaOH	pH
5	2.1	17	8.0
10	2.4	18	8.4
12	2.5	20	9.0
14	2.7	25	9.4
15	3.2	30	9.6
16	5.4	40	9.8

8-4. When the wet sample weight is 1.00 g, what is the capacity of the ion exchanger in question 8-3 above, at the estimated pK value?

8-5. Given the following titration data, estimate the pK value(s), and describe the probable ion exchange material.

Titer, ml of 0.155 N HCl	pH	Titer, ml of 0.155 N HCl	pH
2	7.4	11	4.0
4	6.1	12	3.3
5	5.0	13	2.4
6	4.6	14	2.9
8	4.4	16	1.9
10	4.1	20	1.6

8-6. By titration, an ion exchange sample has a pK value of 5.4. Can this material be used to treat a dilute solution of hydrochloric acid to reduce the acidity? Would it be effective for the removal of acetic acid?

8-7. A water supply contains 75 ppm $Ca(CO)_3$ and 25 ppm $NaHCO_3$ as calcium carbonate, and has a pH of 8.6. Suggest an ion exchange material that can be used to reduce the alkalinity of the water supply and indicate the ion which will appear first at exhaustion.

8-8. Using the pK values listed in Table 4-4, will a weak acid cation exchanger operate efficiently when the water supply has a pH of 3.2?

8-9. Using the pK values listed in Table 4-4, would a weak base anion exchanger operate better at the pH given in question 8-8 above? If so, why?

8-10. What ion exchange material will yield the lowest cost and produce the least spent regenerant waste when used for dealkalization? Give reasons for your choice.

UNIT 9   CATION EXCHANGE CAPACITY

Objective

To determine the total capacity and the salt-splitting capacity of cation exchange materials.

Discussion

The determination of the total capacity of a cation exchange resin defines the number of functional groups on the polymer matrix. Manufacturers use the data, expressed as meq/dry g or meq/ml, in their sales literature. Also, the total capacity is used by the manufacturing facility for quality control of day-to-day production.

The total capacity, expressed as meq/ml, can be used to size industrial equipment when the ion exchange efficiency is known from other data.

Total capacity determinations on used ion exchange materials can verify process problems when the data is compared with that observed on an unused (new) sample of the same lot. The technique employed here involves the exchange of one ion present in large excess in solution for another ion on the ion exchange material. In the case of cation exchange materials, the determination is based on the following reaction:

$$R-SO_3H + \underset{\text{excess}}{NaOH} \rightleftharpoons R-SO_3Na + \underset{\text{excess}}{NaOH}$$

In this case, the NaOH is in excess of the expected ion exchange capacity, and the amount of exchange is found by titrating the residual or excess NaOH with a standardized acid solution.

Salt-splitting, as the following reaction indicates,

$$R-SO_3H + \underset{\text{excess}}{NaCl} \rightleftharpoons R-SO_3Na + HCl + \underset{\text{excess}}{NaCl}$$

is a measure of the functional group strength which splits a neutral salt and forms the corresponding free H^+ ion, which is determined by titration with a standardized base.

Procedure: Total Capacity

Transfer about 10–12 ml of the moist resin sample to a 25 ml graduate, add deionized water, stopper, and invert several times to remove entrapped air. After settling, tap down to minimum volume. Record the tapped-down volume as "A" to the nearest 0.2 ml. Transfer the sample quantitatively to the equipment. Pass 500 ml of 10% HCl downflow through the sample at a flowrate of about 10 ml/min. Rinse the sample with deionized water until free of excess acidity using methyl orange to indicate the completeness of rinse. Discard the spent acid and rinse solutions.

Place a clean 1 L volumetric flask under the collection point, and pass downflow through the sample exactly 500 ml of a previously standarized 0.1 N sodium hydroxide in a 5% NaCl solution at a flowrate of 10 ml/min. Follow the sodium hydroxide solution with sufficient deionized water to fill the volumetric flask to the calibration mark.

Drain the sample completely, transfer the sample to a tared dish or beaker, dry overnight in an oven set at 105°C, cool in a desicator, and record the net dry sample weight as "B" to the nearest 0.01 g.

Mix the contents of the flask and titrate two 50 ml portions with previously standardized 0.1 N sulfuric acid to the phenolphthalein endpoint. Record the average titration as "C" to the nearest 0.1 ml.

Calculation

The total capacity is calculated from the difference between the initial amount of sodium hydroxide and the amount of sodium hydroxide found in the filtrate as follows:

(a) Total capacity by volume:

$$\text{meq/ml} = \frac{[(500 \times 0.1) - (C, \text{ml.} \times N \times H_2SO_4)]}{\text{Sample volume, A, ml}}$$

(b) Total capacity by weight:

$$\text{meq/ml} = \frac{[(500 \times 0.1) - (C, \text{ml.} \times N \times H_2SO_4)]}{\text{Sample dry weight, B, g}}$$

Procedure: Salt-Splitting Capacity

Transfer about 10–12 ml of the sample to a 25 ml graduate, add deionized water, stopper, and invert several times to remove entrapped air. After settling, tap down to a minimum volume and record the minimum volume as "A" to the nearest 0.2 ml.

Transfer the sample quantitatively to the equipment. Pass 500 ml of 10% HCl downflow through the sample at a flowrate of about 10 ml/min. Rinse the sample with deionized water until free of acidity, using methyl orange to indicate the rinse endpoint. Place a clean 1 L volumetric flask under the collection point and pass sufficient 5% NaCl downflow through the sample at a flowrate of 10 ml/min to fill the volumetric flask to the calibration mark.

Rinse the sample with 200 ml of deionized water and discard the rinse water. Transfer the sample quantitatively to a tared dish or beaker, dry overnight in an oven set at 105°C, cool in a desicator, and record the net dry weight of the sample as "B" to the nearest 0.01 g.

Mix the contents of the volumetric flask and titrate two 50 ml portions with previously standardized 0.1 N sodium hydroxide to the phenolphthalein endpoint. Record the average titration as "C" to the nearest 0.1 ml.

Calculation

The salt-splitting capacity is calculated as follows:

(a) Salt-splitting capacity by volume:

$$\text{meq}/\text{ml} = \frac{[C, \text{ml} \times N \times \text{NaOH} \times (1000/50)]}{\text{Sample volume, A, ml}}$$

(b) Salt-splitting capacity by weight:

$$\text{meq}/\text{ml} = \frac{[C, \text{ml.} \times N \times \text{NaOH} \times (1000/50)]}{\text{Sample dry weight, B, g}}$$

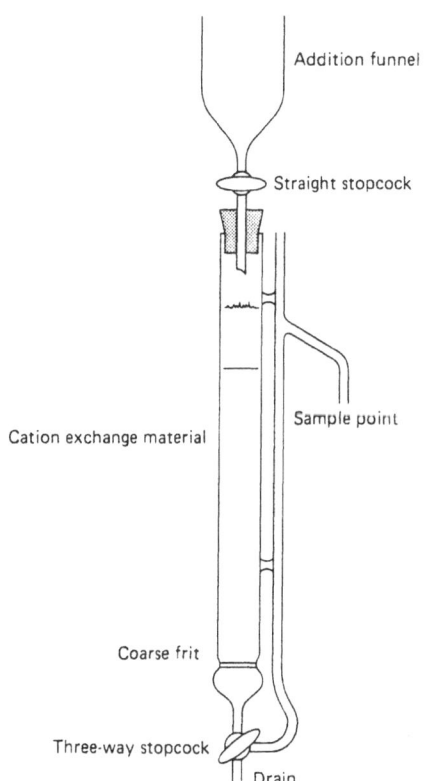

Figure U9-1. Cation exchange capacity.

Your notebook should contain:

—Sample identification
—Normality of all standardized solutions used
—All sample volumes designated ''A''
—All sample weights designated ''B''
—All titration volumes designated ''C''

for each sample of ion exchange material obtained from the laboratory instructor. The above information must be turned in with the answered questions for grading.

QUESTIONS

9-1. Starting with a sample volume of 12.6 ml and a volumetric capacity of 2.24 meq/ml, estimate the average titration volume for the total capacity. And the salt-splitting capacity.

9-2. At 2 meq/ml, what is the maximum sample volume you can use for a total capacity determination? Show calculations.

9-3. If the reactions on an ion exchange material are:

$$M^+ + RcNa \rightleftharpoons RcM + Na$$
$$M^{+2} + 2RcNa \rightleftharpoons RcM + 2Na$$

What will be the pick up of 1 L of resin if M is Silver and M is Copper (Cu)? Assuming the reaction is run to completion and the capacity of the cation exchanger is 2.0 meq/ml, give the weight of the Silver and Copper in grams.

9-4. What does the answer to question 9-3 tell you about the recovery of valuable metals, such as silver, gold, and platinum, by ion exchange.

9-5. Routine cation exchange capacity measurement of an operating unit indicates a 20% loss in 2 years. Give three possible reasons for the observed decrease.

9-6. For each reason stated in question 9-5 above, indicate the test that might be used to verify the cause.

9-7. What results would you expect for the total and salt-splitting capacity of a weak acid cation exchanger $(R-COOH)$? Support your answer with equations.

9-8. Can this procedure be used to determine the capacity of the cation portion of a mixed bed without separation? Explain your answer.

9-9. The volumetric ratio of anion to cation resin of a mixed bed is 1.55. When the cation capacity is 1.96 meq/ml, what will be the total cation volumetric capacity in meq/ml? Show your calculation.

9-10. When the salt-splitting capacity is 97.6% of the total, what is the volumetric salt-splitting capacity in question 9-9 above?

UNIT 10 ANION EXCHANGE CAPACITIES

Objective

To determine the total and salt-splitting capacity of anion exchange materials.

Discussion

Strong base anion exchange materials are generally considered to be mono-functional. However, even strong base anion exchangers, when new, may have some of the total capacity present as weakly ionized (weak base) groups. As these materials age in service, the percentage of weak base groups can increase, and their determination is important. This is particularly true when strong base anion exchangers and strong acid cation exchangers are used in the production of deionized water. The purity of the deionized water produced is dependent upon the strength of the functional group, particularly when weakly ionized electrolytes are present in the water being processed.

When problems arise in the operating performance of industrial equipment, a determination of total capacity of a representative sample may indicate if the trouble is due to the ion exchange material, the equipment, or the mode of operation.

Procedure: Total Capacity

Transfer about 15 ml of the sample supplied by the laboratory instructor to a 25 ml graduate and add deionized water to flood the sample. Stopper the graduate and invert several times to remove entrapped air; allow the sample to settle. Tap down and record the minimum volume as "A" to the nearest 0.2 ml. Transfer the sample quantitatively to the test apparatus being used.

Pass 500 ml of 10% HCl downflow through the sample at a flowrate of about 10 ml/min. Follow this immediately with a 500 ml rinse prepared by mixing 3 volumes of deionized water to 1 volume of isopropyl alcohol. At the end of the alcohol/water rinse, spot check the effluent for Cl^- ion with a few drops of 1% $AgNO_3$. When Cl^- is found, rinse with an additional 500 ml of the mixed alcohol/water. Discard the spent HCl and rinse solutions.

Place a clean 500 ml volumetric flask under the collection point, and pass sufficient 10% $NaNO_3$ downflow through the sample at 10 ml/min to fill the volumetric flask to the calibration mark. Mix the contents of the flask and titrate two 50 ml portions for chloride ion with previously standardized 0.1 N $AgNO_3$. Record the average titration volume as "B" to the nearest 0.1 ml.

Treat the same sample with 500 ml of 10% NaOH, and rinse to a pH of less than 8.0 with a solution prepared by mixing 3 volumes of deionized water with 1 volume of isopropyl alcohol. Discard the spent NaOH and the rinse solutions.

Place a second clean 500 ml volumetric flask under the collection point, and pass sufficient 5% NaCl downflow through the sample at a flowrate of 10 ml/min to fill the volumetric flask to the calibration mark.

Mix the contents of the flask and titrate two 50 ml portions to the phenolphthalein endpoint using previously standardized 0.1 N sulfuric acid. Record the average titration as "C" to the nearest 0.1 ml.

When "C" > 0.1 "B," rinse the sample quickly with 500 ml of deionized water, transfer to a tared dish or beaker, and dry overnight in an oven set at 105°C. Cool in a desicator and record the dry net weight as "D" to the nearest 0.01 g.

When "C" > 0.8 "B," pass 500 ml of 0.1 N HCl through the sample, rinse with 500 ml of deionized water, and transfer to a tared dish or beaker. Dry overnight in an oven set ot 104°C, cool in a desicator, and record the net dry weight as "D" to the nearest 0.01 g.

Calculations

The total and salt-splitting capacity of the anion exchange material is calculated as follows:

(a) Total capacity by volume:

$$\text{meq/ml} = \frac{[\text{B, ml} \times N \times \text{AgNO}_3 \times (500/50)]}{\text{Sample volume, A, ml}}$$

(b) Total capacity by weight:

$$\text{meq/g} = \frac{[\text{B, ml} \times N \times \text{AgNO}_3 \times (500/50)]}{\text{Sample dry weight, D, g}}$$

(c) Salt-splitting capacity by volume:

$$\text{meq/ml} = \frac{[\text{C, ml} \times N \times \text{H}_2\text{SO}_4 \times (500/50)]}{\text{Sample volume, A, ml}}$$

(d) Salt-splitting capacity by weight:

$$\text{meq/g} = \frac{[\text{C, ml} \times N \times \text{H}_2\text{SO}_4 \times (500/50)]}{\text{Sample dry weight, D, g}}$$

The relative percentages of strong and weak base capacity available can be calculated as follows:

Weak base, % = (TC-SSC) × 100/TC
Strong base, % = (SSC × 100)/TC

Your notebook should contain:

—Sample identification
—Normality of all standardized solution used

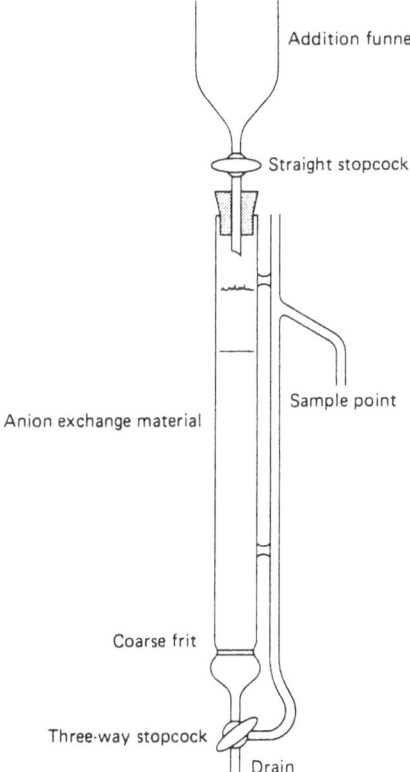

Figure U10-1. Anion exchange capacity.

—All sample volumes designated "A"
—All sample weights designated "D"
—All titration volumes designated "B" and "C"

The above data and the answered questions shall be turned in to the laboratory instructor for grading.

QUESTIONS

10-1. Explain total capacity, strong capacity, and weak capacity?

10-2. If the total capacity is 1.3 meq/ml, and the strong capacity is 1.0 meq/ml, what is the % weak capacity? What effect does the weak base capacity have on silica removal for anion exchange materials?

10-3. If a water supply contains Cl^-, SO_4^{-2}, and $HSiO_2^-$, which ion breaks through first.

10-4. Can a weak base ion exchange material in the regenerated form be used to remove any of the three ions given in question 10-3? Why?

10-5. In what industries and laboratory work is the removal of weak electrolytes of importance.

10-6. Under what operating conditions will a weak base anion exchange material remove chloride and sulfate ions?

10-7. What results would your expect for the total and salt-splitting capacity of a weak base anion exchanger $[R-N-(CH_3)]$? Support your answer with equations.

10-8. Can this procedure be used to determine the capacity of the anion portion of a mixed bed without separation? Explain your answer.

10-9. The volumetric ratio of anion to cation in a mixed bed is 1.55. When the anion capacity is 1.23 meq/ml, what will be the total anion volumetric capacity in meq/ml? Show your calculation.

10-10. When the salt-splitting capacity is 95.4% of the total, what is the volumetric salt-splitting capacity in question 10-9 above?

UNIT 11   SELECTIVITY

Objective

To determine the relative selectivity and the dependence of the selectivity constant on the ion being exchanged.

Discussion

Selectivity measurements are made to indicate the preference of one ion over another by an ion exchange material. This subject and the method for selectivity determination can be found in virtually every book on ion exchange technology.

These data are useful for predicting the behavior of a mixture of ions to be separated by chromatography or the degree of removal of an ion on an ion exchange column. Equilibrium distribution coefficients for the common cation and anions on sulfonic acid and quaternary ammonium exchange materials, respectively, are available in the existing literature.

Chromatographic column separation data for various groups of metallic cations—including the alkali metals, alkaline earths, the transition elements, and the rare earths—are likewise to be found in the literature. The equilibrium behavior of all of the elements in the periodic chart, involving adsorption from solutions of hydrochloric acid on a strong base anion exchange material, has been described.

The relative affinity of cation exchanger materials for the hydrogen ion, and of the anion exchange material for the hydroxyl ion, are of commercial interest in deionization operations. Cation exchangers, having a high affinity for hydrogen ions, require less regenerant; similarly, anion exchangers with a high affinity for hydroxyl ion will require less alkali regenerant. Regenerant economy is favored by a low degree of ionization of the active group on the exchanger matrix. Some variation in regenerant economy may occur with changes in cross-linking.

PART A: CATION SELECTIVITY

Transfer about 20 g of a strong acid cation exchange material to a suitable treatment column (see Unit 6 or 7).

Convert the cation exchanger to the hydrogen form with 500 ml of 10% HCl and rinse with deionized water until the effluent is free of excess acidity as indicated with methyl orange. Drain the resin, dump the converted resin out on a paper towel, blot off excess water, and transfer the sample to a screw cap bottle and seal.

Weigh out 5.898 ± 0.001 g of oven-dried NaCl and transfer to a clean 1 L volumetric flask. Dissolve the salt in deionized water, dilute to the calibration mark with deionized water, mix and label.

Weigh out 5.058 ± 0.001 g of oven-dried CaCl and transfer to a clean 1 L volumetric flask. Dissolve the salt in deionized water, dilute to the calibration mark with deionized water, mix, and label.

From the pretreated ion exchange material weigh out six 3.00 ± 0.01 g portions and transfer each to clean 500 ml Erlenmeyer flasks. Add the following solutions to each labeled flask, and cover each flask with small pieces of paper toweling secured with a rubber band.

Solution	Flask no.	Solution vol (ml)	Deionized water (ml)	Final norm.	meq/ml
0.1 N NaCl	A	50	450	0.010	10
	B	100	400	0.020	20
	C	500	None	0.100	100
0.1 N CaCl	A	50	450	0.010	10
	B	100	400	0.020	20
	C	500	None	0.100	100

Mix each by gently swirling, and allow solutions and ion exchange materials to equilibrate overnight.

Weigh the remaining pretreated ion exchange material and record the weigh as "A" to the nearest 0.01 g. Transfer the material to a 250 ml Erlenmeyer flask, add 100 ml of 10% NaCl, and 4–6 drops of phenolphthalein indicator. Titrate the mixture to the phenolphthalein endpoint with previously standardized 0.1 N NaOH, and record the titration volume as "B" to the nearest 0.1 ml.

Returning to the equilibrating samples, mix each one by gently swirling.

After the resin has settled, analyze each ion exchange/solution mixture as follows:

    a. Transfer a 100 ml portion using a pipette to a clean 250 ml Erlenmeyer flask. Take care that no ion exchange particles are transferred with the solution.

    b. Titrate with previously standardized 0.05 N NaOH to the phenolphthalein endpoint. Record the titration volume as "C" to the nearest 0.1 ml.

Calculations

The following calculations are completed for each equilibration and the final data is added to the appropriate places in the attached report form entitled," Selectivity in Cation Exchange Materials."

    (a) Initial solid phase H ion concentration:

$$\text{H, meq/wet gram} = \frac{\text{"B" ml} \times N \times \text{NaOH}}{\text{"A", g}}$$

    (b) Final solid phase H ion concentration:

$$(\text{H, meq/g})r = \frac{\text{"B" ml} \times N \times \text{NaOH}}{\text{"A", g}} - \frac{\text{"C" ml} \times 0.05 \times (500/100)}{3.00}$$

    (c) Final solid phase Na ion concentration:

$$(\text{Na, meq/wet gram})r = \frac{\text{"C" ml} \times 0.05 \times (500/100)}{3.00}$$

    (d) Final solid phase Ca ion concentration:

$$(\text{Ca, meq/wet gram})r = (\text{Na, meq/wet gram})r$$

    (e) Final aqueous phase H ion concentration:

$$\text{H, meq/ml} = \frac{\text{"C" ml} \times 0.25}{500}$$

    (f) Final aqueous phase Na ion concentration:

$$\text{Na, meq/ml} = \text{Final, meq/ml} - \frac{\text{"C" ml} \times 0.25}{500}$$

    (g) Final aqueous phase Ca ion concentration:

$$\text{Ca, meq/ml} = \text{Na, meq/ml}$$

Calculate the selectivity coefficients for K, using the following equations, and add the final results to the report.

(a) $Na^- + R-SO_3H \rightleftharpoons H^+ + R-SO_3 \cdot Na$

$$K_H{}^{Na} = \frac{(Na, \text{meq/wet gram})r}{(H, \text{meq/wet gram})r} \times \frac{(H, \text{meq/ml})s}{(H, \text{meq/ml})s}$$

(b) $Ca^{+2} + 2R-SO_3H = 2H^- + (R-SO_3)_2Ca$

$$K_H{}^{Ca} = \frac{(Ca, \text{meq/wet gram})r}{(H, \text{meq/wet gram})r} \times \frac{(H, \text{meq/ml})s}{(Ca, \text{meq/ml})s}$$

The selectivity coefficients shown in Table U11-1 have been taken from the *DUOLITE Ion Exchange Manual* for comparison only.

PART B: ANION SELECTIVITY

Weigh out two 15-g portions of the strong base type I. anion exchange material supplied by the laboratory instructor, and transfer each to a suitable apparatus (see Unit 6 or 7). Convert one to the Cl ion form by treating with 500 ml of 10% HCl. Rinse with deionized water until essentially free of excess Cl ion as indicated by a spot test with 1% AgNO solution. Drain the resin sample, dump it out on a paper towel, blot off the excess water, transfer to a screw cap bottle, seal, and label.

Convert the second portion to the SO_4 ion form by treating with 500 ml of 5% H_2SO_4. Rinse with deionized water until essentially free of SO_4 ion as indicated by a spot test using a 1% BaCl solution. Drain the sample, dump it out on a paper towel, blot off the excess water, transfer it to a screw cap bottle, seal, and label.

Determine the initial available capacity of the strong base anion used by employing the pretreated chloride form of the anion exchange material as follows:

(a) Weigh out 3.00 ± 0.01 g of the moist pretreated resin and record the exact weight as "A" to the nearest 0.01 g. Transfer the sample quantitatively to a funnel fitted with a suitable coarse filter paper.

(b) Pass sufficient 10% $NaNO_3$ downflow through the sample to fill a 500 ml volumetric flask to the calibration mark.

(c) Mix the contents of the flask and titrate two 50 ml portions for Cl^- ion with previously standardized 0.1 N $AgNO_3$ solution using potassium dichromate as the indicator. Record the average titration as "B" to the nearest 0.1 ml.

(d) Calculate the initial total capacity as follows:

$$\text{Capac., meq/wet gram} = \frac{\text{``B'' ml} \times N \times \text{AgNO} \times (500/50)}{\text{Sample wet weight, g}}$$

The $Cl-OH$ selectivity coefficient is determined by starting with the strong base anion exchange material in the Cl form.

From the pretreated Cl form anion exchange material weigh out three 3.00 $\pm$ 0.01 portions. Transfer each portion to clean 500 ml flasks. Add the 0.1 N NaOH solution volumes and deionized water volumes as shown below.

Solution	No.	Solution volume (ml)	Deionized water (ml)	Final norm.	meq/ml
0.1 NaOH	A	200	300	0.040	40
	B	400	100	0.080	80
	C	500	None	0.100	100

Mix each sample by gently swirling, cover with a small piece of paper towel held in place with a rubber band, and allow the ion exchanger/solution combination to equilibrate overnight at room temperature. At the end of the equilibration period, mix the samples by gentle swirling and allow the resin to settle. Transfer a 50 ml portion of each solution to clean 250 ml Erlenmeyer flasks using a pipette. Taking care that no ion exchange particles are transferred with the solution. Titrate each aliquot to the phenolphthalein endpoint using previously standardized 0.1 NH_2SO_4. Record the titration volume as "C" to the nearest 0.1 ml.

The SO_4-OH selectivity is determined by starting with the ion exchange material, which has been pretreated to the SO form. Weigh out three 3.00 $\pm$ .01 g portions of the pretreated moist ion exchange resin and transfer each quantitatively to clean 500 ml Erlenmeyer flasks. Add the solution volumes and deionized water volumes as shown in the following table:

Solution	No.	Solution volume (ml)	Deionized water (ml)	Final norm.	meq/ml
0.1 NaOH	A	200	300	0.040	40
	B	400	100	0.080	80
	C	500	None	0.100	100

Mix each sample by gently swirling, cover each with a small piece of paper towel held in place by a rubber band, and allow the ion exchange/solution combinations to equilibrate overnight at room temperature. After equilibration, mix the samples a second time by gently swirling and allow the ion exchange material to settle. Transfer a 50 ml portion of each solution to clean 250 ml Erlenmeyer flasks. Titrate each to the phenolphthalein endpoint using previously standardized 0.1 N H_2SO_4. Record the titration volume of each as "D" to the nearest 0.1 ml.

Calculations

The following calculations are completed for each equilibration, and the final values are written in the appropriate spaces in the report form entitled "Selectivity in Anion Exchange Materials."

(a) Initial solid phase capacity, meq/wet gram (obtained from instructor).

(b) Final solid phase OH ion concentration:

$$H, meq/wet gram = \frac{Final \ OH, \ meq/ml - \text{``B''} \ ml \times N \times AgNO \times (500/500)}{\text{``A,''} \ wet \ grams}$$

(c) Final solid phase Cl ion concentration:

$$Cl, meq/wet gram = \frac{\text{``B''} \ ml \times N \times AgNO \times (500/50)}{\text{``A,''} \ wet \ grams}$$

(d) Final solid phase SO_4 ion concentration:

$$SO, meq/wet gram = \frac{\text{``D''} \ ml \times N \times H_2SO_4 \times (500/50)}{\text{``A,''} \ wet \ grams}$$

(e) Final aqueous phase OH ion concentration:

$$OH, meq/ml = \frac{\text{``C''} \ ml \times N \times H_2SO_4}{50}$$

(f) Final aqueous phase Cl ion concentration:

$$Cl, meq/ml = Final \ OH, \ meq/ml \ \frac{\text{``C,''} \ ml \times N \times H_2SO_4}{50}$$

(g) Final aqueous phase SO ion concentration:

$$SO, meq/ml = Final \ OH, \ meq/ml \ \frac{\text{``D,''} \ ml \times N \times H_2SO_4}{50}$$

Enter all values in the report form and calculate the selectivity coefficients for K with the following equations:

(a) $R-N(CH_3)_3Cl + OH^- = R-N(CH_3)_3OH + Cl^-$

$$K_{OH}{}^{Cl} = \frac{(OH \ meq/wet \ gram)r}{(Cl \ meq/wet \ gram)r} \times \frac{(Cl, \ meq/ml)s}{(OH \ meq/ml)s}$$

Table U11-1. Selectivity Coefficients as a Function of Crosslinking

Ion	4%	8%	12%	16%
		Nominal cross-linking		
		Monovalent species		
H	1.0	1.0	1.0	1.0
Li	0.9	0.85	0.81	0.74
Na	1.3	1.5	1.7	1.9
NH	1.6	1.95	2.3	2.5
K	1.75	2.5	3.05	3.35
Rb	1.9	2.6	3.1	3.4
Cs	2.0	2.7	3.2	3.45
Cu	3.2	5.3	9.5	14.5
Ag	6.0	7.6	12	17
		Divalent species		
Mn	2.2	2.35	2.5	2.7
Mg	2.4	2.5	2.6	2.8
Fe	2.4	2.55	2.7	2.9
Zn	2.6	2.7	2.8	3.0
Co	2.65	2.8	2.9	3.05
Cu	2.7	2.9	3.1	3.6
Cd	2.8	2.95	3.3	3.95
Ni	2.85	3.0	3.1	3.25
Ca	3.4	3.9	4.6	5.8
Sr	3.85	4.95	6.25	8.1
Hg	5.1	7.2	9.7	14
Pb	5.4	7.5	10.1	14.4
Ba	6.15	8.7	11.6	16.54

(b) $\{R-N(CH_3)_3\}_2 SO + 2OH = 2R-N(CH_3)_3 OH + SO_4^{-2}$

$$K_{OH}^{SO_4} = \frac{(OH\ meq/wet\ gram)r}{(SO\ meq/wet\ gram)r} \times \frac{(SO\ meq/ml)s}{(OH\ meq/ml)s}$$

The selectivity coefficients shown in Table U11-2 have been taken from the *DOULITE Ion Exchange Manual* for comparison only.

For approval and grading by the laboratory instructor your notebook should contain all:

—Sample weights used
—Ion exchanger identification
—Solution normalities
—Titration volumes
—All rough calculations

The completed report forms and the answered questions must also be turned in to the instructor for final grading on this unit.

Table U11-2

| Anion | Selectivity coefficient for | |
	Strong Base Type I	Strong Base Type II
OH	1.0	1.0
I	175	17
HSO	85	15
ClO	74	12
NO	65	8
Br	50	6
CN	28	3
HSO	27	3
BrO	27	3
NO	24	3
Cl	22	2.3
HCO	6.0	1.2
IO	5.5	0.5
F	1.6	0.3

QUESTIONS

11-1. Do the values for Na and Ca given in Table U11-1 indicate that the Ca ion can be removed from solution by an Na form ion exchange material? Explain.

11-2. Will an SO_4^{-2} form anion exchanger or a Cl^- form anion exchanger be easier to regenerate with caustic? Why?

11-3. When K = 1.3 and K = 6.0, estimate K.

11-4. If the selectivity for ion M is 3.00, and for ion N it is 6.00, what will happen if a solution having equal concentrations of the two ions is passed through a bed of ion exchange material? Which ion will break through first? What happens when the run is continued beyond the breakthrough?

11-5. Could the behavior described in question 11-4 be used to separate ions M and N?

11-6. Given a water supply containing:

Calcium	70 mg/L
Magnesium	10 mg/L
Sodium	80 mg/L
Lithium	5 mg/L
Potassium	8 mg/L

Indicate (see Table U11-1) the order of appearance in the effluent from a strong acid cation exchange unit.

11-7. Given a water supply containing:

Bicarbonate	105 mg/L
Chloride	11 mg/L
Sulfate	16 mg/L
Silica	8 mg/L
Nitrate	12 mg/L

Indicate (see Table U11-2) the order of appearance in the effluent treated by a type-I strong base anion exchange unit.

11-8. Does the order in question 11-7 change if a type-II strong base material is used instead of the type-I?

11-9. For a type-I strong base anion exchange material, when K = 175 and K = 65; estimate K.

11-10. Estimate K for a type-II strong base anion exchanger.

UNIT 12   RATE OF ION EXCHANGE

Objective

To determine the effect of particle size on the reaction rate of the ion exchange process.

Discussion

The determination of an ion exchange reaction rate is of practical significance since the velocity of flow used in a column will be somewhat dependent on the rate of reaction. Comparison of the relative reaction rates of different ion exchange materials with the same polar group is also helpful in evaluation of relative porosities. In general, the reaction rate for a given process depends upon the diffusion of the ions in and out of the ion exchange particle. As a rule, the diffusion rate is limited by either the liquid film around the particle or by the particle itself. Equations have been developed to describe these diffusion phenomena. The literature indicates that the reaction rate is increased by:

1. Increasing the agitation to reduce the film thickness.
2. Increasing the ion concentration to augment the driving force.
3. Decreasing the particle size to increase the surface area and to decrease the effect of solid diffusion.
4. Increasing the temperature to increase the ion diffusion in both the film and the solid phases.

The following experimental procedure is specifically focused upon the effect of particle size on the exchange of sodium ion in solution for hydrogen ion in the ion exchange material.

Procedure

Weigh out six 5.00 ± 0.01 g portions of a moist hydrogen form strong acid cation exchange material from a prescreened $-16 + 20$ mesh fraction. Transfer each sample quantitatively to individual clean 500 ml Erlenmeyer flasks labeled 5, 10, 20, 30, 60, and 120 min, respectively. As rapidly as possible add 500 ml of 0.5% NaCl to each flask, starting with the 5-min test. Record the time of the first addition.

Mix each flask by swirling, allow the ion exchange material to settle, and at each appropriate time decant about 100 ml of the NaCl solution into individual clean beakers properly labeled. Take care that no ion exchange particles are transferred with the solution.

Titrate a 50 ml aliquot of each with a previously standardized 0.05 N NaOH solution to the phenolphthalein endpoint. Record the titration volume of each as ''A'' to the nearest 0.1 ml. Record the contact times and the corresponding titration volumes in a suitable table.

Repeat the same experiment starting with the same ion exchange material using a prescreened $-40 + 50$ mesh fraction.

Calculations

Calculate the capacity of each sample as meq/wet gram, and plot as a function of contact time as shown in the attached example. Obtain the value for infinite contact time (equilibrium value) from the laboratory instructor for the ion exchange material used.

Assuming a first-order reaction rate, the amount of H^+ ion exchanged per unit time can be expressed as an exponential equation: $N_t = e^{-Kt}$, where $N_t = (meq/g)_{equal.} - (meq/g)_{at\ anytime}$. Therefore, when the values for N_t, are plotted on semilog paper as a function of time, the data should yield a unique straight line for each mesh fraction used. The rate constant, K, for each particle size can be determined by calculating the slope of the straight line.

Your notebook should contain the following information when completed:

—All sample weights
—Sample identification
—Room temperature check
—Solution normalities
—Titration volumes
—Graphs and calculations

A completed report sheet, and the answered questions, must be turned in to the instructor with your notebook for grading.

Rate of ion exchange

Particle size Fraction	Contact Time (min)	Titration volume (ml)	Capacity (meq H/wet gram)
−16 + 20	5	_____	_____
	10	_____	_____
	20	_____	_____
	30	_____	_____
	60	_____	_____
	120	_____	_____
−40 + 50	5	_____	_____
	10	_____	_____
	20	_____	_____
	30	_____	_____
	60	_____	_____
	120	_____	_____

$$\text{meq. H/wet gram} = \frac{A_{120} \times N \times \text{NaOH} \times (500/50)}{\text{Wet sample weight, g}}$$

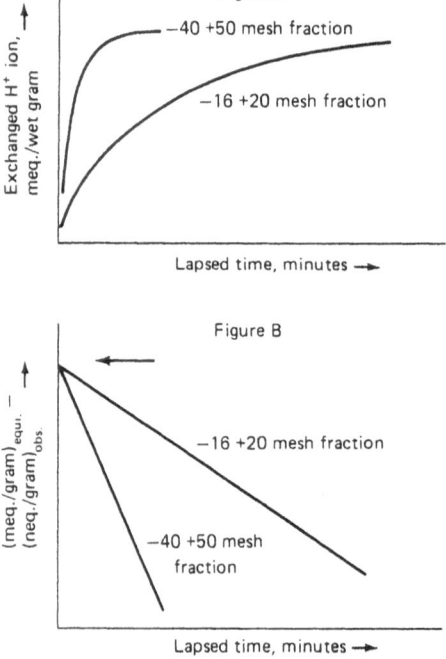

Figure A

−40 +50 mesh fraction

−16 +20 mesh fraction

Exchanged H⁺ ion, meq./wet gram

Lapsed time, minutes →

Figure B

−16 +20 mesh fraction

−40 +50 mesh fraction

(meq./gram)equi. − (neq./gram)obs

Lapsed time, minutes →

Figure U12-1. Rate of exchange.

QUESTIONS

12-1. Does the viscosity of water have any effect on the ion exchange rate?

12-2. Knowing that the size of the beads may vary from lot to lot, could it have an effect on the uniformity of test results?

12-3. By consulting a book on ion exchange technology, define:

Particle diffusion _____

diffusion _____

12-4. How does concentration of the anion in solution determine which of the two types of diffusion is of greater importance?

12-5. How would the results differ if a macroporous ion exchange material were used?

12-6. Does the ion exchange reaction rate have a direct bearing on leakage levels in certain applications?

12-7. Give at least two other examples which may cause ion leakage.

12-8. Will high flowrates have a positive or negative effect on the ion exchange process? Explain.

12-9. Can a low reaction rate be compensated for by using a deeper ion exchange bed?

12-10. Explain the advantages of using finer particle sizes and shallower beds.

UNIT 13 STABILITY OF ION EXCHANGE MATERIALS

Objective

To determine the stability of a strong base anion exchange material in the hydroxide form to exposure to high-temperature water.

Discussion

When the use of an ion exchanger is contemplated for any application, prediction of resin life is always a major consideration. Quick tests which might be used in making such predictions would be most valuable. Unfortunately, experience has shown that accurate estimates of service life cannot be made on the basis of simple accelerated tests since there are many variables involved.

Nevertheless, a few such tests have been used by which comparisons can be made. Any predictions resulting from these tests are necessarily relative rather than absolute, but they have been useful. Results, for example, have been used to predict relative service life of two analogous ion exchange materials in the same application. With very few exceptions, field results have qualitatively confirmed the experimental data.

The laboratory tests used are both specific and general. The specific tests are carried out by contacting the candidate resin with the actual solutions to be processed. Such experiments can be either static batch contacts or scaled-down column studies of the relative effects. The latter technique is preferred, since it simulates the conditions under which the ion exchange material will be used. Rapid cycling of small columns with the process solutions can provide useful information. Raising the temperature in such experiments accelerates attainment of data in accordance with the well-known Van't Hoff law.

The more general tests are based upon the effects which are known to cause ion exchanger degradation, such as heat, oxidation, and irreversible adsorption (fouling). The suitability of an amine anion exchange material for beet, cane, or corn sugar deionization can be evaluated by exposure of the resin to the effects of dextrose at elevated temperatures. Exposure of an ion exchange material to a high temperature for a short period of time, and determination of the effects of the exposure on the chemical and physical properties, often yields significant comparative information.

The conditions to be employed vary considerably with the ion exchanger under test. In general, most significant results are obtained by exposing the resin in its least stable form. Thus, anion exchangers are tested in the hydroxyl form and cation exchangers are tested in the hydrogen form. Of all the ion exchange materials available, strong base anion exchangers in the hydroxyl form have the least stability. Heat causes the conversion of the quaternary ammonium groups to tertiary amine groups, and there may be some nitrogen loss. Therefore, simple thermal stability tests are most commonly employed with strong base anion exchange materials, even though most of the service cycle is at ambient temperature. To improve regeneration efficiency and get better silica removal, strong base anion exchangers are frequently regenerated with NaOH at temperatures as high as 120°F (49°C). In such applications, temperatures in the resin bed approach levels at which the ion exchanger begins to exhibit instability. Thermal stability tests are made on cation exchange materials only when the application involves temperatures above 300°C (150°C), such as in hot lime processes followed by ion exchange softening.

In addition to stability in high-temperature environments, ion exchange materials are sensitive, in varying degrees, to:

—Oxidizing systems
—Mechanical attrition
—Osmotic shock

Procedure

This procedure is written as a general method. It is suggested that at least three different ion exchange materials should be evaluated for stability to bring out the relative differences between ion exchanger types.

Transfer about 30 g of the as-received moist ion exchange material to a filter crucible with a coarse frit. Treat the sample with 500 ml of 10% HCl at a flowrate of about 15 ml/min. Rinse the sample with deionized water until the filtrate is free of acidity as indicated by methyl orange. Drain the sample by suction, weigh out 5.00-0.01 g and transfer it quantitatively to a clean 250 ml Erlenmeyer flask labeled with the appropriate ionic form. Dry a second 5.00-0.01 g sample in an oven set at 105°C overnight, cool in a desiccator, and record the net dry weight as "A" to the nearest 0.01 g.

Treat the remaining sample with 500 ml of 5% NaOH solution and rinse free of excess NaOH with deionized water using phenolphthalein to indicate completeness of rinse. Drain the sample by suction, weigh out 5.00-0.01 g and transfer it quantitatively to a clean 250 ml Erlenmeyer flask labeled with the appropriate ionic form. Dry a second sample of the treated ion exchanger in an oven at 105°C overnight, cool in a desiccator, and record the dry weight as "B" to the nearest 0.01 g.

To each flask containing the ion exchange material in a known ionic form, add 100 ml of 3% hydrogen peroxide solution and 1 ml of ferrous sulfate solution containing 11.9 g of $FeSO_4-7H_2O/L$. Stopper each flask with glass wool plugs and place both on a water bath regulated at 65 3°C (150 3°F) for 6 hr.

Transfer one resin sample and solution quantitatively to a tared filter crucible and wash with 200 ml of hot (65 3°C) deionized water. Wash with 100 ml of acetone. Repeat the filtration procedure with the second ion exchange sample. Dry both samples in an oven set at 105°C overnight, cool in a desiccator, and record the net dry weight as "C" to the nearest 0.01 g for the sample treated with HCl and as "D" for the sample treated with NaOH.

Calculations

The degree of oxidative degradation is calculated for each sample and for each ionic form as follows:

For the HCl treated sample: % Degradation = $(A - C) \times 100/A$
For the NaOH treated sample: % Degradation = $(B - C) \times 100/B$

In order to receive a grade for this unit, you must answer all the questions and have the instructor review your complete notebook work.

QUESTIONS

13-1. Why are type-II ion exchange materials kept in the salt form during storage even when in closed containers?

13-2. What is the difference between strong base type-I, strong base type-II, and weak base anion exchange materials as to temperature stability?

13-3. In a strong acid cation exchange material, the styrene–divinylbenzene matrix is the least resistant component. In anion exchange materials the weakest component is?

13-4. At high temperatures, what does the anion group in a type-II strong base resin break down to?

13-5. Why is regeneration of a type-I anion exchanger done at 60°C? What happens to silica removal after a high-temperature regeneration?

13-6. What is the first property of a cation exchange material that degrades under oxidizing conditions?

13-7. What is the first property of an anion exchange material that degrades under oxidizing conditions?

13-8. Describe osmotic shock and its major effect on ion exchange materials.

13-9. Ion exchange products are sometimes specifically manufactured to be resistant to osmotic or mechanical shock. Will such products have higher or lower operating capacities than conventional products of the same type? Explain why.

13-10. What is the major effect of continuous systems on ion exchange materials?

UNIT 14   COLUMN OPERATING CAPACITY OF ION EXCHANGERS

PART A: CATION EXCHANGERS

Objective

To demonstrate the differences in the operating characteristics of strong acid and weak acid cation exchange materials.

Discussions

The operating capacity and characteristics of a cation exchange material for a given application is probably the most significant measurement that can be made. When the synthetic influent water is a reasonably close approximation of the actual raw water supply, the data obtained can be translated into an industrial scale unit operation. In addition, regenerant efficiency and water quality can be predicted with reasonable certainty. The technique is often used for application outside the field of water treatment (i.e., special applications).

Procedure: Strong Acid Cation Exchangers

Transfer 72 ± 0.1 g of the moist as-received strong acid cation exchange material to a test column measuring 1.9 cm inside diameter by 50 cm long (Figure

U14-1). Use deionized water to transfer the material quantitatively and drain the excess water to the top of the resin bed. Determine and record the B&D bed volume by following the procedure in Unit 3.

Prepare 12 L of a test water with the following composition using deionized water.

Anions	Content (%)	Cations	Content (%)
HCO_3	70	Na	40
SO_4	10	Mg	30
Cl	20	Ca	30

An example of the technique used to prepare synthetic influent waters is shown on the attached calculation sheet. Regenerate the cation exchanger with 510 ml of 2% H_2SO_4 (dosage = 10 lb/ft^3 or 160 kg/m^3). Pass the solution downflow at 10 ml/min. When the acid solution has reached the top of the ion exchanger bed, add 300 ml of deionized water and continue at the same flowrate until the rinse water reaches the top of the resin bed.

Introduce the test water and adjust the flowrate to 40 ml/min, collecting the treated effluent in 250 ml portions. Each portion will be analyzed for the following:

(a) Acidity, meq/L: Titrate a 50 ml aliquot with previously standardized 0.02 N NaOH to the methyl red endpoint—1 ml titer = 0.4 meq/L acidity.
(b) Total Hardness, meq/ml: Titrate a 50 ml aliquot with previously standardized 0.02 N sodium-EDTA to the Eriochrome Black endpoint—1 ml titer = 0.4 meq/L total hardness.
(c) Effluent sodium ion leakage can be determined by atomic absorption or ion selective electrode when the equipment is available.

The column operation is continued uninterrupted at the flowrate of 20 ml/min until the effluent acidity exhibits a 20% drop from the average value observed during the middle part of the exhaustion run. Keep a plot of the effluent acidity and total hardness as a function of the volume of water treated.

Procedure: Weak Acid Cation Exchangers

This test method uses the same synthetic water as given in the previous section. Transfer 64 g of the moist as-received weak acid cation resin, supplied by the laboratory instructor, to a column measuring 1.9 cm inside diameter by 50 cm long (Figure U14-1). Use deionized water to quantitatively transfer the ion exchange material and drain the excess water to the top of the resin bed. Determine and record the B&D bed volume by following the procedure in Unit 3.

Regenerate the ion exchange material with 700 ml of 0.8% H_2SO_4 (dosage = 4 lb/ft^3 or 64 kg/m^3). Pass the solution downflow through the column of resin at a flowrate of 6 ml/min. When the acid solution reaches the top of the bed, add 300 ml of deionized water and continue at the same flowrate until the rinse water also reaches the top of the ion exchange bed.

Introduce the test water, adjust the flowrate to 40 ml/min and collect the effluent in 250 ml portions. Each portion will be analyzed for the following:

(a) Acidity, meq/L: Titrate a 50 ml aliquot of each with previously standardized 0.02 N NaOH to the methyl red endpoint—1 ml titer = 0.4 meq/L.

(b) Total hardness, meq/L: Titrate a 50 ml aliquot of each with previously standardized 0.02 N sodium-EDTA to the Eriochrome Black endpoint—1 ml titer = 0.4 meq/L.

(c) Alkalinity, meq/L: Titrate a 50 ml aliquot of each with previously standardized 0.02 N H_2SO to a methyl red endpoint—1 ml titer = 0.4 meq/L.

(d) Effluent sodium ion can be determined by atomic absorption or ion selective electrode when the equipment is available.

Continue the column operation at 20 ml/min uninterrupted until the effluent alkalinity expressed as meq/L is more than 10% of the test water alkalinity titrated in the same way.

Calculations

The uncorrected operating capacity of the ion exchange materials is calculated as follows:

$$\text{Capac., meq/ml} = \frac{\text{Vol., L} \times \text{Conc., meq/L}}{\text{Bed Volume, ml}}$$

When the data is available, the operating capacity can be corrected for cation leakage using the following expression:

$$\text{Capac., meq/ml} = \frac{\text{Vol., L} \times (\text{Conc., meq/L} - \text{Average Leakage, meq/L})}{\text{Bed Volume, ml}}$$

PART B: ANION EXCHANGERS

Objective

To demonstrate the relative differences in the operating characteristics of strong base and weak base anion exchange materials.

Discussion

The ion exchange operating capacity and characteristics of an anion exchange material for a given application is probably the most important measurement that can be made. When the influent water used is a close approximation of the actual raw water supply, then the data observed can be translated into full-scale industrial equipment. In addition, the information obtained is useful for predicting the water quality, regenerant efficiency, and leakage levels to be expected.

Procedure: Strong Base Anion Exchangers

Prepare 12 L of test water for the evaluation of anion exchange operating capacity as follows:

(a) Fill a suitable container with 10 L of deionized water. Add 6 ml of concentrated hydrochloric acid and 3.5 ± 0.1 g of concentrated sulfuric acid.

(b) In a separate container, dissolve 300 mg of sodium metasilicate in about 500 ml of deionized water. This solution is passed quickly through a 100 ml bed of strong acid cation exchanger in the hydrogen form. The effluent is added immediately to the acidic solution from step (a), and mixed.

(c) The solution is diluted to 12 L with deionized water, mixed, and analyzed for acidity, Cl ion concentration, and silica concentration as SiO_2.

Transfer 55.0 ± 0.1 g of the as-received moist strong base anion exchanger provided by the instructor to a test column measuring 1.9 cm inside diameter by 50 cm long (see Figure U14-1). Use deionized water to transfer the ion exchange material quantitatively, and drain the excess water down to the top of the bed. Determine and record the B&D volume by following the procedure in Unit 3.

Regenerate the anion exchange material with 320 ml of 4% NaOH (dosage = 10 lb/ft^3 or 160 kg/m^3. Pass the solution downflow through the ion exchange bed at about 3 ml/min. When the caustic solution reaches the top of the resin bed, add 300 ml of deionized water and continue at the same flowrate until the rinse water reaches the top of the ion exchanger bed.

Introduce the test water, and adjust the flowrate to 40 ml/min. Collect the effluent in 250 ml portions. Each portion will be analyzed for the following:

(a) Effluent conductivity expressed as microohms/cm
(b) Effluent pH
(c) Effluent silica colorimetrically

Continue the column operation at the flowrate of 40 ml/min until the effluent conductivity exhibits a rapid increase, indicating a breakthrough of anions. Using interpolation if necessary, estimate the effluent volume when the conductivity is 20 microohms/cm, and record the volume as "A" to the nearest 0.1 L.

Procedure: Weak Base Anion Exchangers

Transfer about 50 ± 0.1 g of the weak base anion exchange material supplied by the instructor to a test column measuring 1.9 cm inside diameter by 50 cm long (see Figure U14-1). Use deionized water to quantitatively transfer the material and drain the excess water to the top of the resin bed.

Regenerate the column with 150 ml of 4% NaOH solution (dosage = 4 lb/ft^3 or 64 kg/m^3). Pass the solution downflow through the ion exchange material at a flowrate of 5 ml/min. When the caustic solution reaches the top of the bed, add 300 ml of deionized water and continue at the same flowrate until the rinse water reaches the top of the ion exchanger bed.

Introduce the test water and adjust the flowrate to 40 ml/min. Collect the effluent in 250 ml portions. Each portion will be analyzed for the following:

(a) Effluent silica colorimetrically
(b) Effluent conductivity expressed as microohms/cm.

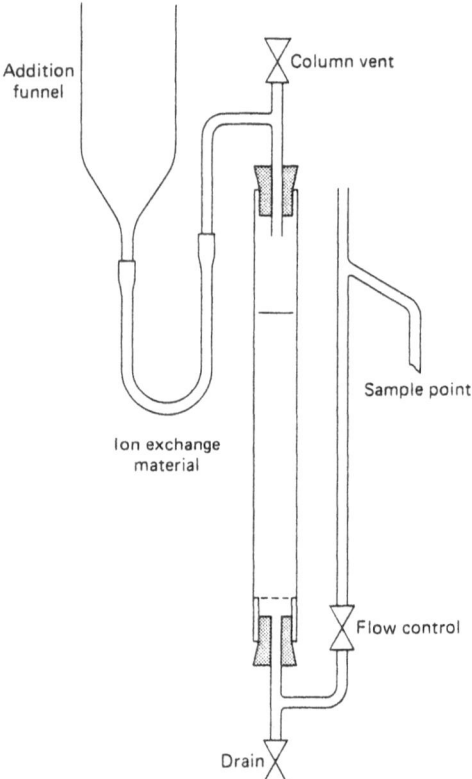

Figure U14-1. Column operating capacity. Part A: Cation exchangers. Part B: Anion exchangers.

Continue the column operation until the effluent conductivity exhibits a rapid rise, indicating a breakthrough of anions. Using interpolation if necessary, estimate the effluent volume when the conductivity is 20 microohms/cm and record the volume as "A" to the nearest 0.1 L.

Calculations

The operating capacity of anion exchange materials for acids is calculated as follows:

$$\text{Capacity, meq/ml} = \frac{A \times \text{Conc. Meq/ml}}{\text{Bed Vol. ml}}$$

where, for strong base resins, TA = (acidity + silica) as mg/L. For weak base resins, TA = acidity as mg/L.

Your notebook should contain all bed volumes, resin identifications, weights used, and the analytical data for the test waters prepared for this unit. The laboratory notebook records and the answered question must be turned in to the instructor for grading when this experimental unit is completed.

QUESTIONS

Part A: Cation Exchangers

14-1. When used in a two-bed demineralizer system, which ion exchange material is the best choice? Give reasons for your answer.

14-2. In a given decationization application, a strong acid cation exchanger yields 650 gal/ft^3 (86840 L/m^3) of bed. The water supply is changed to a new supply which contains 445 ppm as $CaCO_3$ as compared to 220 ppm as $CaCO_3$ for the old well. What will the treated water volume be on a 75 ft^3 bed between regenerations?

14-3. What happens to the bicarbonate ion on passing through a hydrogen-form cation exchange material?

14-4. What process can be used to remove the by-products formed in question 14-3?

Part B: Anion Exchangers

14-5. What is meant by leakage during the service or exhaustion run?

14-6. Will the leakage increase with the increase of ionic concentration of the influent water? Why?

14-7. Will leakage increase with an increase in regenerant dosage?

14-8. Which part of an ion exchanger bed is more fully regenerated, the top or the bottom? And why?

14-9. What size particles are on the bottom compared to those on the top of the bed? What effect does this have on the rates of reaction? Can you relate these factors to leakage?

14-10. Why is the leakage reduced when counterflow is used in contrast to co-flow?

UNIT 15   DEIONIZATION

Objective

To demonstrate the difference between two-bed, and mixed-bed deionization on the same test water.

Discussion

The terms "demineralization" and "deionization" have come into common use and are synonyms. They are used to identify the ion exchange process used to remove dissolved ionic constituents from water supplies. Briefly, the deionization process is carried out by removing the cations with an acid-form cation exchange material followed by the removal of the anions with a hydroxyl-form anion exchange material. The overall process will result in the production of water free on ions—hence, the term "deionization."

The first deionization process used in industry was a two-bed system which employed a strong acid cation exchanger followed by a weak base anion exchanger. The reactions involved in this process are as follows.

Cation exchange:

$$
\begin{array}{l}
NaCl \\
CaCl_2 \\
MgSO_4
\end{array}
\quad + R{-}SO_3H \rightleftharpoons
\begin{array}{l}
-Na \\
R{-}SO_3{-}Ca \\
-Mg
\end{array}
\quad +
\begin{array}{l}
H_2SO_4 \\
HCl
\end{array}
$$

Acid absorption:

$$
R{-}NH_2 + HCl = R{-}NH_2{-}HCl
$$
$$
H_2SO_4 \qquad -H_2SO_4
$$

Deionization systems of this kind do not remove weakly ionized species from solution such as $HSiOb_2{}^-$, since it exists in solution as H_2SiO_3. As industry increased its demand for pure water, removal of silica became more important and this was achieved by using strong base ion exchange materials. The reactions involved are as follows.

Cation exchange:

$$\text{CaCl} \qquad\qquad -\text{Ca} \qquad \text{HCl}$$

$$\text{Na SiO} + \text{R}-\text{SO}_3\text{H} = \text{R}-\text{SO}-\text{Na} + \text{H SiO}$$

$$\text{MgSO} \qquad\qquad -\text{Mg} \qquad \text{H}_2\text{SO}_4$$

Anion exchange

$$\text{HCl} \qquad\quad -\text{Cl}$$

$$\text{R}-(\text{CH}_3)_3\text{OH} + \text{H}_2\text{SiO}_3 = \text{R}-(\text{CH}_3)_3\text{SiO} + \text{H}_2\text{O}_4$$

$$\text{H}_2\text{SO}_4 \qquad -\text{SO}_4$$

Finally, the systems evolved, where the two ion exchange resins (cation and anion) were put in the same unit and operated as a mixed-bed. Now industrial deionization systems come in a great variety of combinations as shown in Figure U15-1. The system used, of course, depends on the water supply and the final water quality required.

Procedure: Two-Bed Deionization

Arrange two test columns in series as shown in Figure U15-1. Weigh out 62 ± 0.01 g of H^+ form strong acid cation exchange material and transfer it quantitatively into the first column using deionized water. Weigh out 51 ± 0.01 g of hydroxyl-form strong base anion exchange material and transfer it quantitatively to the second column using deionized water. Carefully fill all free spaces with deionized water. Add 500 ml of deionized water to the addition funnel. Pass the deionized water through the two-bed system and vent any accumulated air. Allow the deionized water to drain until it reaches the addition funnel stopcock, and stop flow.

Weigh out 5.84 ± 0.01 g of oven-dried NaCl, dissolve in deionized water, and dilute to 10 L with deionized water. This test water contains 10 meq/L and is used for both the evaluation of two-bed deionization and the evaluation of the mixed bed.

Add the test water to the addition funnel, start flow through both beds in series, and adjust the flowrate to 30 ml/min. Keeping the funnel filled with test water at all times, measure and record the effluent conductivity (microohms/cm) every 400 ml. When the conductivity starts increasing, record the conductivity readings more frequently, stop when the conductivity has passed 50 microohms/cm.

Plot the recorded conductivity values as a function of effluent volume and determine the volume treated to a 10 microohms/cm endpoint using interpolation if necessary.

Record the volume as "V" to the nearest 0.1 L.

Determine the B&D cation exchanger bed volume and record it as "C" to the nearest whole milliliter.

Determine the B&D anion exchanger bed volume and record it as "A" to the nearest whole milliliter.

Calculations

The capacity of each ion exchange bed is calculated independently as follows.

For the cation column:

$$\text{Capacity, meq/ml} = (V \times 10)/C$$

For the anion column:

$$\text{Capacity, meq/ml} = (V \times 10)/A$$

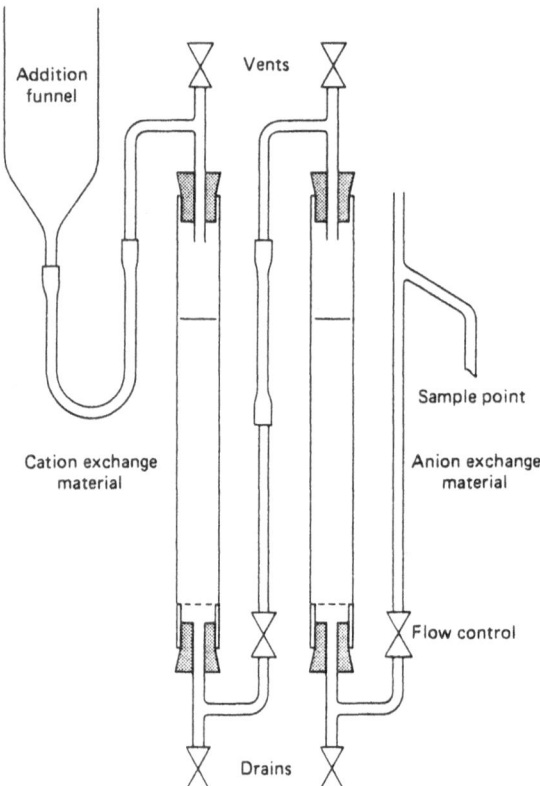

Figure U15-1.   Deionization. Part A: Two-bed treatment.

Procedure: Mixed-Bed Deionization

Weigh out 60 ± 0.1 g of the as-received moist mixed-bed ion exchange material supplied by the instructor. Using as little deionized water as possible, transfer the material quantitatively to the test column, taking great care that the mixed anion and cation components *do not* separate. Add enough deionized water to displace the air from the top of the mixed bed. Attach the addition funnel, and fill with the prepared 10 meq/ml/NaCl solution.

Start flow through the mixed bed and adjust the flowrate to 30 ml/min. Keep the addition funnel filled with the test solution at all times, and record the effluent conductivity (microohms/cm) every 200 ml. Follow the same procedure used for the two-bed test and again stop when the effluent conductivity is greater than or equal to 50 microohms/cm. Plot the conductivity readings as a function of effluent volume, and determine the effluent volume treated at 10 microohms/cm by interpolation if necessary. Record the effluent volume treated

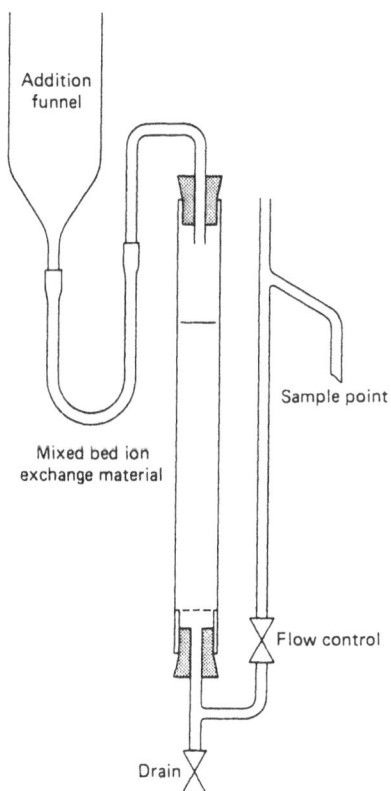

Figure U15-2. Deionization. Part B: Mixed-bed system.

as "V" to the nearest 0.1 L. Determine the B&D volume of the exhausted mixed-bed and record as "B" to the nearest whole milliliter.

Calculations

The capacity of the mixed bed is calculated as follows:

$$\text{Capacity, meq/ml} = (V \times 10)/B$$

A copy of the experimental data including all weights and plotted curves should be turned in to the instructor with the answered questions for grading.

QUESTIONS

15-1. What system tested produces the purest effluent in terms of residual total dissolve solids (electrolytes)? Why does this happen?

15-2. In a two-bed system how can one tell which unit exhausted first?

15-3. Why do most industrial water treatment systems use a sequence consisting of: cation unit, degasifier, anion unit, and mixed bed?

15-4. If a water supply to be treated contains Ca, Mg, Na, HCO_3 (e.g., 50 ppm) Cl, SO_4, and SiO_3 (e.g., 5 ppm), which unit in question 15-3 has the heaviest load? Which is next, and which has the lightest load?

15-5. How is a mixed bed regenerated?

15-6. Calculate the process volume, in gal/ft^3, of a cation exchange material with an operating capacity of 18.2 kg/ft^3 for an influent water containing 55 mg/L as CaCO?

15-7. If the influent is 30% bicarbonate and a degasifier is used, what is the process volume of an anion exchange material with a rated capacity of 12.6 kg/ft^3?

15-8. If the leakage from the two-bed system described in questions 15-6 and 15-7 above is 1% of the influent, what is the process volume, in gal/ft^3 of the final mixed bed with a capacity of 12.3 kg/ft^3?

15-9. Above a few hundred ppm total dissolved solids, mixed-bed demineralization can be uneconomical. Describe two systems which might be used to reduce operating costs.

15-10. Can regenerant recycle (reuse) be used to further reduce both costs and waste? Explain.

UNIT 16   CHROMATOGRAPHY

Objective

To demonstrate the effectiveness of ion exchange materials for the chromatographic separation of anions and cations in aqueous media.

Discussion

Ion exchange chromatography is a method used for the separation of ions on an ion exchange material by virtue of the different affinity the ion exchanger has for different ions. It is an important analytical tool and has led to many interesting results in inorganic, organic, and biochemistry. From the standpoint of the different forms of chromatography, ion exchange chromatography is a liquid–solid technique in which the ion exchange material represents the solid phase. The separation of ions in ion exchange chromatography can be based on ion selectivity, ion exclusion, or ion retardation phenomena of the ion exchange materials. As the ions are selectively absorbed, certain ions will be displaced in a consistent manner depending upon the position of the equilibrium prevailing between the concentrations in solution and the resin phase and—to a lesser degree—on the temperature. For the present, all types of ion exchange material have been used in ion exchange chromatography. The technology is now employing ion exchange materials with very low capacity (i.e., 0.02–0.05 meq/ml) to obtain high resolution of closely related ions. The techniques used to separate ions by ion exchange chromatography fall into four groups:

- Frontal Chromatography
- Displacement Chromatography
- Elution Chromatography
- Development with Organic Solutions

The simplest procedure is frontal chromatography. It consists of the continuous injection of the solution, which contains the components that are to be separated at the head of the column. The individual components are chromatographically separated in order of their increasing selectivity while moving through the column. The appearance of individual ions in the effluent is followed by any suitable analytical method.

In displacement chromatography, the ions to be separated are charged to the upper end of the ion exchange column. The separation is achieved by eluting with an electrolyte solution which contains ions that are preferentially absorbed by the ion exchanger, so that the ions of interest are displaced down the column.

As an example, it will be assumed that the ion exchanger was in the hydrogen form and that sodium and potassium were to be separated, with the selection of calcium chloride solution as the displacement medium as shown in Figure U16-1. The selectivity of the ions involved is H < Na < K < Ca (i.e., the first counter-ion is weakly bound and the one serving for displacement is strongly bound). The chromatogram develops as a result of the displacement of all other counter-ions from the uppermost layers of the column by the calcium ions and their advances in front of the latter in the course of development of the chromatogram with the formation of a sharp front. Depending on their relative selectivity H, Na, and K ions are bound more or less strongly by the exchanger, and as a result they will fall into a specific order with self-sharpening fronts. The ions to be separated will appear in the eluate in order to their selectivity. A certain degree of zone-overlapping cannot be avoided.

This method provides pure and mixed fractions, with the latter containing the two ions which are adjacent to each other in terms of their selectivity sequence. When desirable, the mixed fractions can be purified further by repeating the entire operation. This method is, therefore, mainly used for preparative separations.

In elution chromatography, the ions to be separated are charged to the top of the column as in displacement chromatography and are exchanged for the counter-ion of the ion exchange material on the top of the column. For the chromatographic separation, the column is treated with an eluant containing the same ion which initially forms the counter-ion of the entire exchanger. Although the eluant with its counter-ion overtakes the bound ions, a separation occurs during the course of the elution. Thus, the sites become occupied by the counter-ion of the eluant, which is the counter-ion of the original ion exchanger. The ions separate into zones in the course of their migration down the column and emerge from the column together with the co-ion of the eluant.

Due to its special nature, the cost of equipment, and the size limitations, this experiment may be handled as a demonstration by the instructor. When class participation is possible, it is suggested that the experiment be carried out by class teams of several individuals assigned by the instructor.

Procedure: Anion Separations

Prepare the following solution:

Stock Solution

> Weigh out 1.272 ± 0.001 g of dry NaCl
> Weigh out 1.545 ± 0.001 g of dry Na_2SO
> Weigh out 1.189 ± 0.001 g of dry Na_3PO

Test Solution.   Transfer these salts quantitatively to a clean 500 ml volumetric flask, then dissolve and dilute to the calibration mark with deionized water. One milliliter of this solution contains 1000 mg chloride, 1000 mg sulfate, and 1000 mg phosphate.

This stock solution is used to prepare a test solution in the following way. Transfer a 1 ml portion of the stock solution to a clean 100 ml volumetric flask, and dilute to the calibration mark with deionized water. This working solution has the following concentrations of the ions of interest.

- 10 mg/L as chloride
- 10 mg/L as sulfate
- 10 mg/L as phosphate

Eluant Solution.   Weigh out 0.386 ± 0.001 g of m-phenylenediamine hydrochloride and transfer it quantitatively to a clean 1 L volumetric flask with deionized water. Using a pipette, transfer 25 ml of 0.1 N hydrochloric acid to the volumetric flask and dilute to the calibration mark with deionized water. This eluant solution contains:.

0.0025 M HCl
0.0025 M m-phenylenediamine hydrochloride

Fill the reservoir with eluant, start pump, and adjust flowrate to 2 ml/min. Use the first 3 min of operation to establish a base line on the conductivity recorder. With a Hamilton Syringe, inject exactly 1 ml of the test solution without stopping the pump, and record the time of injection as $t = 0$. Continue pump operation and keep the reservoir filled with the eluant, until three distinct peaks have been recorded. These peaks should be labeled 1, 2, and 3, respectively. Using appropriate spot tests, identify the predominant anion present in each sample collected at each peak. Label and date the conductivity recording, and identify each peak observed.

Procedure: Cation Separations

Prepare the following solutions:

Stock Solution

Weigh out 0.758 ± 0.001 g of dry BaCl
Weigh out 0.905 ± 0.001 g of dry SrCl
Weigh out 1.384 ± 0.001 g of dry CaCl

Transfer these salts quantitatively to a clean 500 ml volumetric flask, dissolve and dilute to the calibration mark with deionized water. One milliliter of this solution contains 1000 mg barium, 1000 mg strontium, and 1000 mg calcium.

Test Solution.   Using a pipette, transfer 1 ml of the stock solution to a clean 100 ml volumetric flask and dilute to the calibration mark with deionized water.

Eluant Solution

Weigh out 0.252 ± 0.001 g of dry $NaHCO_3$
Weigh out 0.212 ± 0.001 g of dry Na_2CO_3.

Transfer these salts quantitatively to a clean 1 L volumetric flask, dissolve and dilute to the calibration mark with deionized water. This eluant solution contains:

- 0.003 M sodium bicarbonate
- 0.002 M sodium carbonate

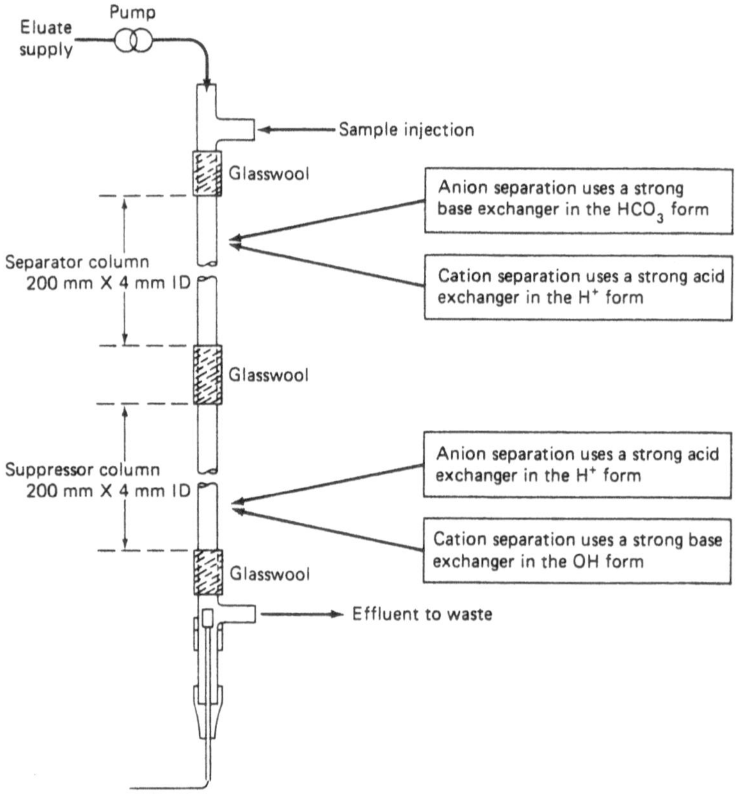

Pump
Eluate supply
Sample injection
Glasswool
Anion separation uses a strong base exchanger in the HCO_3 form
Cation separation uses a strong acid exchanger in the H^+ form
Separator column 200 mm X 4 mm ID
Glasswool
Anion separation uses a strong acid exchanger in the H^+ form
Cation separation uses a strong base exchanger in the OH form
Suppressor column 200 mm X 4 mm ID
Glasswool
Effluent to waste

Figure U16-1. Chromatography.

Height
Width
Base line
Retention time

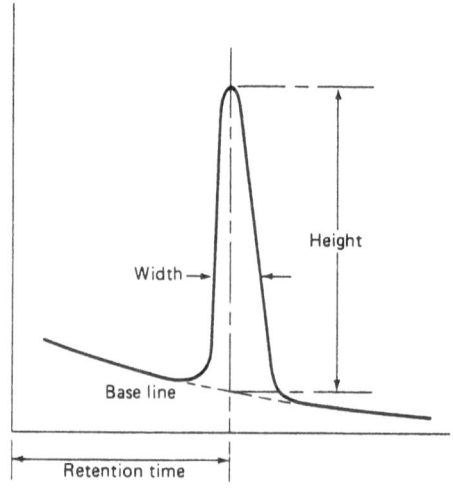

Figure U16-2. Chromatography. Example of conductivity response.

Fill the reservoir with the eluant solution, start pump, and adjust the flowrate to 2 ml/min. Use the first 3 min of operation to establish a base line on the conductivity recorder. Using a Hamilton Syringe, inject 1 ml of the test water at the top of the column without stopping the pump, and record the time of injection as $t = 0$. Continue the pump operation and keep the reservoir filled until three distinct peaks have been recorded. Label the peaks 1, 2, and 3, respectively. Using appropriate spot tests, identify the predominant cation present in each peak. Label and date the conductivity recording and identify each peak.

Calculations

Using the conductivity recording chart, determine the retention time (in minutes) of each identifiable peak. Calculate the area under the peak as follows:

Area, $mm^2 = W \times H$
Where W = Width at half the peak height, mm
H = Peak height above the base line, mm

In order to obtain a grade, the team must turn in the conductivity charts properly labeled. In addition, each student should turn in their independent answers to the questions below.

QUESTIONS

16-1. Can one predict the order of appearance of ions from their selectivities? Explain.

16-2. How would you separate two ions which have small differences in the relative selectivity coefficient?

16-3. What other properties of an eluant may be utilized for separating ions on an ion exchange material? Describe how each technique might be utilized.

16-4. Given a water supply with Ca, Mn, K, and Ag, refer to UNIT 11 and write the relative order of elution.

16-5. What is the relative order of elution of the following halogens Br, F, Cl, and I from a type-II strong base anion exchange material?

16-6. Describe frontal chromatography.

16-7. Why are small ion exchange particles preferred for chromatography applications?

16-8. Can nonionic compounds be separated and analyzed by ion exchange chromatographic techniques?

16-9. Describe elution chromatography.

16-10. What other class of compounds can be separated by ion exchange chromatography?

UNIT 17   ION EXCLUSION

Objective

To demonstrate the ability of ion exchange materials to separate ionic and non-ionic species without regeneration.

Discussion

The ion exclusion process was developed in the early 1950s. It is a process for separation of strong electrolytes from weak electrolytes and nonionic species by using ion exchange materials without regeneration. No actual ion exchange occurs, since the exchange material acts only as a sorbent for the nonelectrolytes. The experimental unit used here will work equally well when the column is filled with either strong acid cation or strong base anion exchange materials.

Procedure

Weigh out 4.00 $\pm$ 0.01 g of dry NaCl in a clean 100 ml beaker, add 40.0 + 0.1 g of glycerin, 50 ml of deionized water, and mix. Titrate a 5 ml portion of the clear solution for chloride ion using previously standardized 0.1 N AgNO$_3$ to the potassium dichromate endpoint and record the titration volume as "A" to the nearest 0.1 ml. Using an Abbe refractometer, measure and record the refractive index of the clear solution and determine the % glycerin concentration (Figure U17-1).

Using a suitable test column, add enough strong acid cation exchange material (Na form) to fill the column about 75%. Rinse with a small amount of deionized water, and drain the excess deionized water to the top of the resin bed. With a suitable pipette, transfer 25 ml of the NaCl/glycerin solution to the ion exchange column and drain enough water from the column to get all the solution into the top.

Note: 25 ml = (25 × 4)/50 = 2.00 g NaCl
            = (25 × 40)/50 = 20.0 g glycerin

Attach addition funnel, add 300 ml of deionized water, start flow, and adjust flowrate immediately to 3 ml/min. Collect the effluent in 20 ml portions and analyze each portion as follows:

- Titrate a 5 ml portion for chloride ion using the same previously standardized 0.1 N AgNO solution and record the titration volumes to the nearest 0.1 ml.
- Measure the refractive index of each effluent sample and record the concentration of glycerin as % from Figure U17-2.

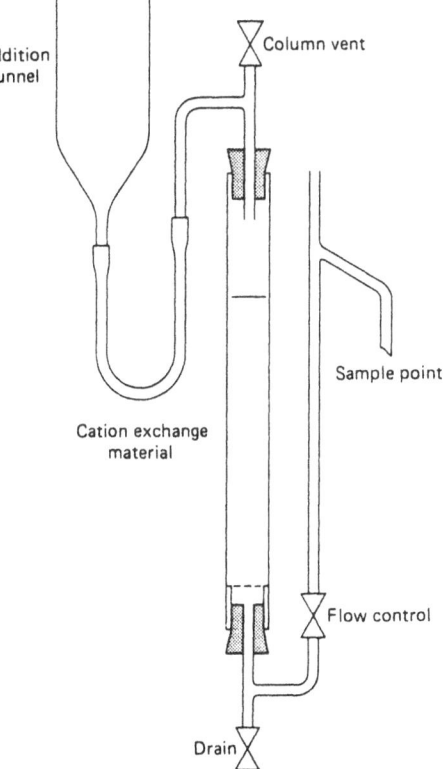

Figure U17-1. Ion exclusion.

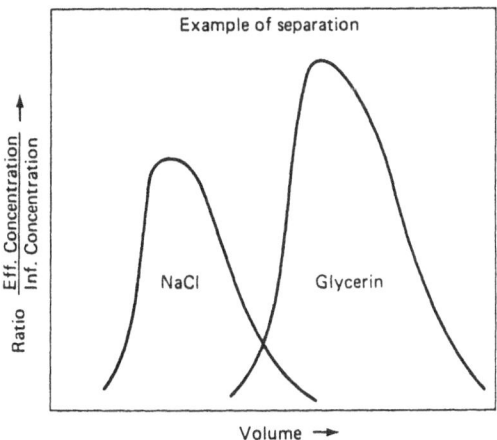

Figure U17-2. Ion exclusion. Example of separation.

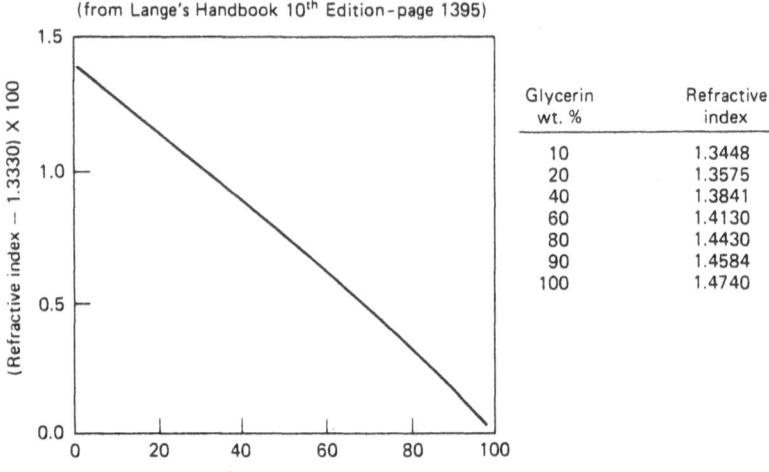

Refractive index versus glycerin concentration (from Lange's Handbook 10th Edition - page 1395)

Glycerin wt. %	Refractive index
10	1.3448
20	1.3575
40	1.3841
60	1.4130
80	1.4430
90	1.4584
100	1.4740

Figure U17-3. Ion exclusion. Refractive index versus glycerin concentration. (From *Lange's Handbook*, 10th ed., p. 1395.)

Continue the column operation until the glycerin concentration in the samples is $\leq 1\%$.

Calculations

For each of the samples, calculate the following ratios:

- Cl ion ratio = Sample titer, ml/"A," ml
- Glycerin ratio = Sample wt%/original wt%

Plot each of these ratios as a function of effluent volume. From Figure U17-3, determine the volume cuts which would yield solutions containing only NaCl and only glycerin.

Turn in the answered questions, and the experimental data to the instructor for grading on this unit.

QUESTIONS

17-1. Could one separate acetic acid from glycerol by using an ion exchange column? Why?

17-2. Is the experiment performed in this unit a means of purifying glycerin?

17-3. Why is water an eluant for glycerin at the end of the cycle?

17-4. How does ion exclusion differ from chromatography? (Review unit 16.)

17-5. Give at least three nonionic compounds that might be separated by ion exclusion.

17-6. What are the primary properties of the dissolved electrolytes that allows the process of ion exclusion to work?

17-7. Would high cross-linked ion exchangers be better for ion exclusion than low cross-linked products of the same type?

17-8. Given a mixed solution of HCl and CH_3COOH, can separation be achieved by ion exclusion? Outline the process.

17-9. What ion exchange product might be used to separate NaCl and NH—Cl?

17-10. What ion exchange product might be used to separate HCl from C—H—COOH?

UNIT 18   ADSORPTION

Objective

To demonstrate the difference between the adsorptive properties of a gel-type and macroporous-type weak base anion exchange material.

Discussion

Many ion exchange materials, particularly anion exchangers, are frequently required to function as adsorbents for organic species in water supplies. In fact, some anion exchange materials are commonly used as organic "scavengers," "traps," "screens," etc., before the raw water is demineralized by conventional processes. In order to function properly in this kind of application, the anion exchanger used must be porous and must have exchange groups with a high degree of polarity. Relative adsorption data is useful for both new and used ion exchange materials. Weak base anion exchangers should be in the free base form, and free of excess moisture when applied. For good comparative measurements, samples must have the same approximate particle size distribution.

Procedure

Prepare a standard dye solution by dissolving 100 mg of Green Dye No. 3 (available from H. Kohnstamm & Co., New York) in 100 ml of 0.1 N hydrochloric acid. This solution should be freshly prepared and must be discarded if more than 8 hr old.

Using a colorimeter (B&L Spectronic 20 or equivalent), adjust the dye solution so that a 2:3 volumetric dilution with additional 0.1 N hydrochloric acid yields an optical density reading of 0.72 ± 0.10 at a wavelength of 420 nm. To determine the optical density of the standard dye solution, multiply the reading of the dilute solution by 2.5. Transfer 100 ml of the adjusted dye solution to a clean 200 ml beaker, add a magnetic stirrer bar, place on a stirring plate, and begin mixing gently. Weigh out 0.50 ± 0.01 g of dry anion exchanger. Add the dry resin sample to the dye in the beaker and record the time of addition as $t = 0$.

Every 15 min, stop stirrer, transfer 2 ml of the dye solution to a colorimeter cell, and record the optical density at 420 nm wavelength. The optical density readings should be done as rapidly as possible, and repeated every 15 min until a total of at least six such reading are made. When evaluating a poor adsorbent, it may be necessary to dilute the 15- and 30-min readings to obtain accurate optical density values. Generally, a dilution of $1:1$ with 0.1 normal hydrochloric acid is sufficient, in which case the true optical density is equal to two times the instrument readings. Record the optical density readings in table form for each period of contact expressed in minutes.

Calculations

For each optical density observation made calculate the % adsorption as follows:

- Let D_0 = optical density of original dye solution
- Let D_t = optical density of samples taken at different times

$$\text{Adsorption, \%} = (D_0 = D_t) \times 100/D_0$$

This experiment can be carried out with several different types of anion exchange materials. It is suggested that a wide variety of dry anion exchange products be made available to the class for this experimental unit.

Your notebook should contain all calibration data, weights, sample identifications, and tabulated readings. This information and the answered questions must be turned in to the instructor in order to receive a grade on this unit.

QUESTIONS

18-1. Is the experiment given in this unit adsorption or ion exchange? Give reasons for your answer.

18-2. From the structure of the dye, can a relation be equated between it and the ion exchanger structure? active group? or active groups?

18-3. What is the molecular weight of the dye? What can you say about the size compared to Cl^- or SO_2^{-2} ions?

18-4. Knowing the size of various ionic species, could one use the information to estimate the pore size of an ion exchange material? Explain.

18-5. Since adsorption is a surface-related process, are porous products more efficient than gel-type products? Explain why.

18-6. What two organic compounds often found in surface water supplies are adsorbed on anion exchange materials during operation?

18-7. Considering the molecular weight of the compounds in question 18-6, would you suspect that the adsorption is irreversible?

18-8. Describe a technique that might be useful for cleaning an organically fouled anion exchange material.

18-9. Would porous ion exchange products be easier to clean, in general, than gel-type products of the same kind? Why?

18-10. Does the nature of the co-ion on the resin matrix play a role in the adsorption process?

UNIT 19   ION EXCHANGE CATALYSIS

Objective

To demonstrate the effectiveness of a hydrogen-form strong acid cation exchange material as a solid catalyst.

Discussion

The catalytic properties of ion exchange materials have been known for many years. One good example is the solid exchange catalyst used in the petroleum industry to crack crudes into lower boiling-point species to aid separations and increase yields. The importance of the use of the solid ion exchange catalyst arises from the fact that it can be readily separated from the products formed. Ion exchangers can be viewed as solids which have the properties of acids or bases. It is not surprising than that they act as catalysts since many organic reactions will go to completion under the influence of free acids and free bases. The main limitation reported is the ability of the reactants to reach the active exchange sites. In general, for flowing reactors, the diffusion rate in the particle has been the controlling factor. Under these conditions, the reaction efficiency is found to decrease as the square root of the particle size, and activation energy was found to be dependent on particle size.

Much attention has been devoted to reactions which are catalyzed by hydrogen-form cation exchange materials. Some of those which have been studied in detail are:

- Inversion of sucrose
- Hydrolysis of esters
- Dehydration
- Epoxidation

Anion exchange materials in the hydroxyl form have been applied to a number of reactions also; some of these are:

- Aldol condensation
- Methanolysis
- Condensations
- Reductions

Ion exchangers have also demonstrated the ability to control and reduce side reactions, prevent color formation, and in general provide good preparative procedures with no contamination of the products with free acids or bases. The ion exchange materials are often fouled by the reactants and may break down in some extreme applications. In addition, they have a definite thermal stability problem when applied at high temperatures.

It is evident from diffusion characteristics that a resin with a high surface area should be a good catalyst. The development of macroporous ion exchange materials enabled them to perform as catalysts in nonaqueous media, where such reactions were very slow and inefficient with conventional gel resins. Ion exchange materials offer the following advantages in catalytic applications:

- The catalyst can be easily separated from the reaction products.
- The product remains free from impurities.
- High yields are possible since side effects are minimized due to the short contact time.
- The catalysis can be controlled to proceed in a specific direction because of the variable properties of the ion exchange material used.
- Ion exchangers can be used repeatedly by regeneration, or by incorporation into continuous processes.

Procedure

To a suitable flask (Figure U19-1) attached to a reflux condenser, add 62 ± 0.1 g of ethylene glycol and 120 ± 1 g of glacial acetic acid (be careful when handling the glacial acetic acid).

Place the flask in a water bath, start the magnetic stirrer, and record the starting time as $t = 0$. At 15-min intervals, remove 1 ml samples of the mixture and titrate each immediately with previously standardized 0.1 N NaOH to the phenolphthalein endpoint. Record each sample time and its corresponding titration volume to the nearest 0.1 ml in an appropriate table. Continue sampling for at least 3 hr.

Repeat the experiment with the same chemicals and weights, and add 2.0 ± 0.1 g of dry hydrogen-form strong acid cation exchange material at the start

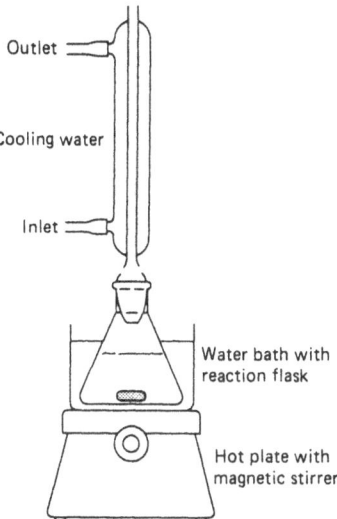

Figure 19-1. Catalysis.

of the test period. Carry out the same sampling procedure and add the observed titration volumes to the table of data.

Plot the titration volumes as a function of time and draw a smooth curve through the data points. Submit the tabulated data, the data plot, and the answered questions to the instructor for grading.

QUESTIONS

19-1. Why are ion exchange materials generally required to be in the H^+ or OH^- form when used as catalysts?

19-2. What is meant by catalyst fouling?

19-3. If a solution contains ferrous ions and a strong base anion exchanger is used as a catalyst, what will happen to the catalyst?

19-4. If the resin was a cation exchanger in question 19-3, what would happen to its effectiveness as a catalyst?

19-5. List three catalytic processes that use strong acid cation exchange materials.

19-6. What two advantages do ion exchange materials have over other catalytic processes?

19-7. What one ion exchange property limits their usefulness in industrial applications?

19-8. List three catalytic processes that use strong base anion exchange materials.

19-9. Since ion exchange catalysis is, in general, a surface reaction, are porous or gel-type products more efficient and why?

19-10. Does the co-ion on the resin matrix take part in the catalytic reaction of interest?

UNIT 20   ION EXCHANGE PLATE HEIGHT

Objective

To demonstrate the effect of flowrate variations on the measurement of the ion exchange plate height, and to review other factors which effect the plate height.

Discussion

The concept of an ion exchange plate evolved from chemical engineering where equivalent plates or transfer units are commonly used to design packed towers for distillation separations.

In ion exchange technology, the height of a theoretical exchange plate, HTEP, is dependent on particle size, solid diffusion, film diffusion, and lateral diffusion. In very simple terms, these variables can be related as follows, since the effects are additive:

$$\text{HTEP} = H_p + H_s + H_f + H_l$$

Then by definition in the aqueous phase:

H_p is the contribution due to the particle size.
H_s is the contribution due to the solid diffusion.
H_f is the contribution due to the film diffusion.
H_l is the contribution due to lateral diffusion.

A rigorous treatment of ion exchange data from many applications yielded the following equation, which accounts for the above effects and relates them to the known physical-chemical properties of ion exchange materials.

$$\text{HETP} = 1.64 r_0 + \frac{\lambda_i}{(\lambda_i - \beta)^2} \times \frac{0.14 r_0^2 v}{D_s} + \left(\frac{\lambda_i}{\lambda_i - \beta}\right)^2$$

$$\times \frac{0.266 r_0^2 v}{D_a(\pi 70 r_0 v)} + \frac{D_a \beta \sqrt{2}}{v}$$

The correspondence between the initial simple expression and this rigorous treatment is given below:

$$H_p = 1.64 \times r$$

$$H_s = \frac{\lambda_i}{(\lambda_i - \beta)^2} \times \frac{0.142 \times r_0^2 \times v}{D_s}$$

$$H_f = \left(\frac{\lambda_i}{\lambda_i - \beta}\right)^2 \times \frac{0.226 \times r_0^2 \times v}{D_a \times [(1 + (70 \times r \times v)]}$$

$$H_l = \frac{D_s \times B \times \sqrt{242}}{v}$$

where

r_0 = the mean particle radius (cm)

v = the unrestricted fluid velocity in the empty column (cm/sec)

D_s = the solid phase diffusion rate (cm^2/sec)

D_a = the aqueous phase diffusion rate (cm^2/sec)

β = the dimensionless void volume of the bed = $0.325 + 1.479 \times (D_p/D_c)$

$\lambda = \beta$ as indicated below

and where D_p is the mean particle diameter (cm) and D_c is the column inside diameter (cm).

This experimental unit uses a very favorable reaction

$$2R-Na + Ca^{2+} \rightleftharpoons R_2-Ca + 2Na$$

and the value of λ_i is assumed to be 8.

The evaluation of HTEP from basic parameters, however, is difficult, since many of the values needed may not be available for a specific application. Fortunately, there are experimental techniques which can be used to estimate the height of a practical exchange plate (HPEP).

When the concentration history, C/C_0, is plotted against the volume of effluent treated, a curve similar to that shown in Figure U20-1 is obtained. The height of the exchange plate can be derived from this data in the following way:

Let

V_e = effluent volume when C/C_0 = 0.05 ml

V_t = effluent volume when C/C_0 = 0.95 ml

$V_z = (V_t - V_e)$ The effluent volume for zone development (ml)

X = operating flowrate (ml/min)

The time required for the ion exchange zone to form and move its own length down a column is:

$$\theta_z = \frac{(V_t - V_e)}{X} = \frac{V_z}{X} \quad \text{expressed in min}$$

and the total time θ_t is equal to V_t/X, which again is expressed in minutes.

The height equivalent of a theoretical exchange plate, HETP, and the height of a practical exchange plate, HPEP, have the following applications:

- They are useful for calculating the bed depth required to achieve a predetermined effluent purity. This is particularly helpful when the HPEP is related to actual operating conditions.
- It is often used to determine, or estimate parameters (i.e., flowrate, column length, particle size, etc.) required for good chromatographic separations.

In the first case, the bed depth calculated can aid the basic design of a treatment by defining the number of anion or cation units required to satisfy the effluent purity requirements.

In the second case, when the value of λ_i can be determined independently for each component in a mixture, the other parameters can be varied to optimize the separation, since H_s, H_f, and r will also effect the HETP.

Defining Q as the area over the curve between $C/C_0 = 0.05$ and $C/C_0 = 0.95$, and Q_{max} as the area bounded by V_e, V_t, $C/C_0 = 0.05$ and $C/C_0 = 0.95$, a factor, F, is calculated as follows:

$$F = Q/Q_{max} \quad \text{and} \quad \theta_f = \theta_z \times (1 - F)$$

For symmetrical curves, F will be equal to 0.5. However, this is rarely the case, and F must be defined for each application of interest. The areas required can be readily obtained by applying Simpsons rule. Another technique for obtaining the areas of interest with good accuracy is given in the experimental procedure.

The time, θ_t, required for the development of the complete breakthrough curve is given by:

$$\theta_t = V_t/X$$

Finally, using the column depth, h, in centimeters, the plate height, unique for the application under study, can be calculated as follows:

$$\text{HPEP} = h \times [\theta_z/(\theta_t - \theta_f)]$$

Procedure

From the as-received moist Na-form strong acid cation exchange material, available from the instructor, weigh out about 70 g. With dionized water transfer the resin to the test column measuring 1.9 cm inside diameter by at least 50 cm long (Figure U20-2).

Determine the bed depth several times using the technique for B&D volume given in Unit 3. Record the average bed volume as ''h'' to the nearest 1.0 ml. Drain the excess deionized water to the top of the ion exchange bed.

Prepare a test water by dissolving 27.75 ± 0.01 g of dry reagent grade

$CaCl_2$ in deionized water and diluting to 25 L with deionized water. Determine the total hardness of the test water at least three times, and record the average as "C_0" to the nearest whole ppm (mg/L).

Using deionized water, set the flowrate at 15 ml/min. Quickly drain the excess deionized water to the top of the bed and introduce the test water without changing the flowrate. Collect the effluent in 250 ml portions and label each sample in sequence. Continue operation without changing the flowrate until 12 L of test solution have been passed through the column.

Analyze each sample for total hardness and calculate the value of C/C_0. Plot the data on cross-section graph paper as a function of effluent volume treated, and construct a smooth curve through the data points.

Repeat the experiment with a fresh bed of cation exchange material from the same lot at a flowrate of 175 ml/min.

Calculations

Do the following calculations and enter the data in the suggested report form (Figure U20-3) to determine the HPEP.

(a) Record the bed depth to the nearest whole millimeter.

(b) Record the operating flowrate to the nearest ml/min.

(c) By interpolation, if necessary, record the effluent volume when C/C_0 = 0.05, as V_e to the nearest 10 ml.

(d) By interpolation, if necessary, record the effluent volume when C/C_0 = 0.95, as, V_e, to the nearest 10 ml.

(e) Calculate $V_z = V_t - V_t$, and record the value to the nearest 10 ml.

(f) Calculate $\theta_t = V_t/X$, and record the value to the nearest 0.1 min.

(g) Calculate $\theta_z = V_z/X$, and record the value to the nearest 0.1 min.

(h) Cut out the square portion of each graph between V_e, V_t, C/C_0 = 0.05 and C/C_0 = 0.95. Weigh and record the weight as A nearest 0.01 g.

(i) Cut out the area over the curve for each graph, and weigh and record the weight as B to the nearest 0.01 g.

(j) Calculate the ratio $F = B/A$ and enter the result to the nearest 0.001 unit.

(k) Complete the computation for HPEP for each test condition and report the value to the nearest 0.1 cm.

Your notebook should contain all data and calculations for the averages used. In addition, clearly identify the ion exchange material used and describe the conditions which apply for each test (i.e., flowrate, particle size, etc.). The laboratory instructor will use your notebook and the answered questions for grading this unit.

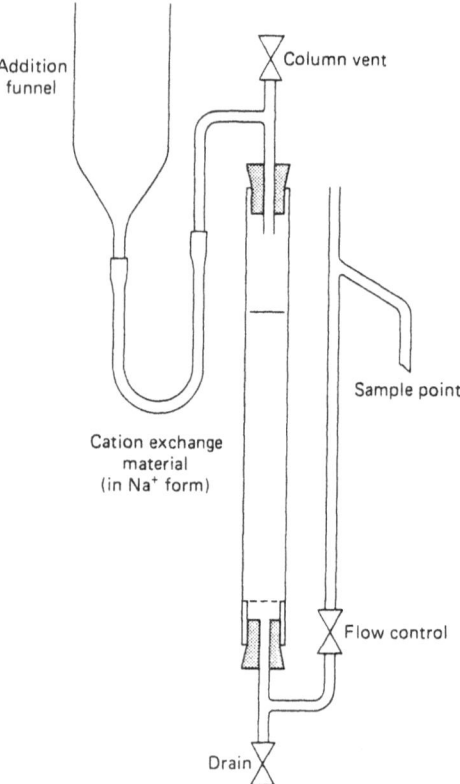

Figure U20-1.   Ion exchange plate height.

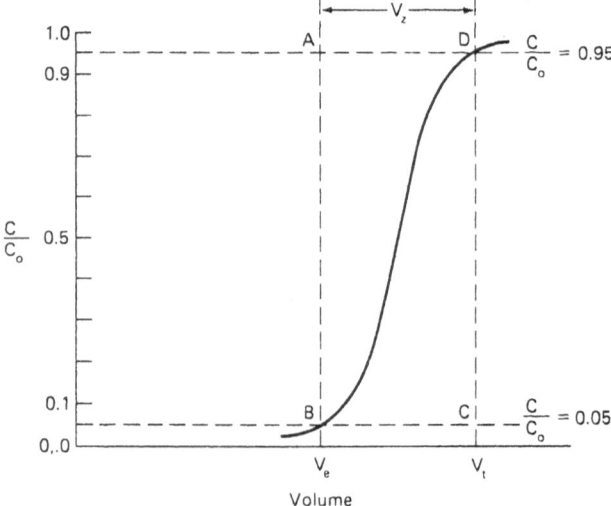

Figure U20-2.   Ion exchange plate height. Typical breakthrough curve.

Test No.	A	B
Bed Depth, h, CMS		
Flow Rate, x, ml/mm		
Eff. Volume when $C/C_0 = 0.05$, Ve, ml		
Eff. Volume when $C/C_0 = 0.95$, V_i, ml		
Calculate $V_z = V_i - V_e$, ml		
Calculate $\theta_i = V_i/x$, min		
Calculate $\theta_z = V_z/x$, min		
Weight of area bounded by V_e, V_i, $C/C_0 = 0.05$ and $C/C_0 = 0.95$, A		
Weight of area over the curve, B		
Calculate $F = B/A$		
Calculate $\theta_f = \theta_z \times (1 - F)$, min		
HPEP, cms $= h\left(\dfrac{\theta_z}{\theta_i - \theta_f}\right)$		

Figure U20-3. Suggested report form.

QUESTIONS

20-1. A reduction of particle size from 0.4–0.04 mm will increase or decrease the number of exchange plates in a given bed volume? Show mathematical solution.

20-2. Given that the relationship between the effluent concentration, C_e, and the influent concentration, C_i, for a single plate is:

$$\frac{C_e}{C_i} = e^{-n}$$

where n is the number of plates. How many exchange plates are required to reduce an influent concentration of 100 mg/L to 0.05 mg/L?

20-3. When a single exchange plate is 7.05 cm, how deep is the bed in problem 20-2? Is this a practical bed depth or shall two beds be used in series.

20-4. What basic changes need to be made to conventional ion exchange equipment to provide the high resolution and sharp breakthrough for analytical chromatography?

20-5. For most mixed-bed applications the HTEP is small and uniform. Explain why this is so.

20-6. Will a twofold increase in flowrate decrease or increase the HPEP? Explain why.

20-7. Given the following data:

$$\text{Flowrate} = 20 \text{ ml/min}$$
$$\text{bed depth} = 30 \text{ in.}$$

Treated volume (L)	C/C_0
1	0.01
2	0.01
4	0.02
4.5	0.18
4.7	0.38
4.9	0.41
5.1	0.50
5.2	0.59
5.3	0.68
5.5	0.84
6	0.97
8	0.98
10	1.00

Estimate the HPEP and show all graphs, weights, and calculations.

20-8. Given that the HPEP is 6.55 cm and a required reduction factor of 20 is necessary, how deep should the ion exchange bed be?

20-9. Condensate polishers often operate at flowrates of 50 gal/ft^2 or more. How does this effect the HTEP?

20-10. What percent reduction or increase occurs if the flowrate is increased a factor of 10 or 20?

APPENDIXES

Appendix A
SUGGESTED READING LIST

For those who wish to become better acquainted with the field of ion exchange technology, the following reading list is offered. Those publications marked with (a), are to the writers knowledge *out of print*; however, they should be available in a technical library. The publications marked with (b) are still in print and available from the publishers as of January 1989. Finally, those publications marked with (c) may be available from International University Microfilm at Ann Arbor, Michigan.

1(c) Amphlett, C. B., *Inorganic Ion Exchangers*, Elsevier. New York (1964).

2(b) Applebaum, S. B., *Demineralization by Ion Exchange*, Academic Press, New York (1968).

3(b) Calmon, C., and Gold, H. (eds.), *Ion Exchange for Pollution Control*, CRC Press Boca Raton, Florida (1979).

4(c) Calmon, C., and Kressman, T. R. E. (eds.), *Ion Exchangers in Organic and Biochemistry*, Interscience Publishers, New York (1957).

5(a) Dorfner, K., *Ion Exchange: Properties and Applications*, Ann Arbor Science Publ. Inc., Ann Arbor, Michigan (1972).

6(c) *Discussions of Chromatographic Analysis*, Faraday Society (1949).

7(c) Glasstone, S., *An Introduction to Electrochemistry*, D. Van Nostrand, New York (1942).

8(c) Helfferich, F., *Ion Exchange*, McGraw-Hill, New York, (1962).

9(c) Inczedy, I., *Analytical Applications of Ion Exchange*, Pergammon Press, New York (1966).

10(c) *Ion Exchange and Its Applications*, Society of Chemical Industries (1954).

11(a) Kelley, W. P., *Cation Exchange in Soils*, Reinhold Publ., New York (1948).

12(a) Khym, J. X., *Analytical Ion Exchange Procedures in Chemistry and Biology*, Prentice-Hall, Englewood Cliffs, New Jersey (1947).

13(c) Kunin, R., *Elements of Ion Exchange*, Reinhold Publ., New York (1960).

14(b) Kunin, R., *Ion Exchange Resins*, R. E. Krieger Publ., Huntington, New York (1971).

15(b) McInnis, D., *Principles of Electrochemistry*, Reinhold Publ., New York (1942).

16(b) Mantell, C. L., *Industrial Electrochemistry*, 1st ed., McGraw-Hill, New York (1931).

17(b) Marinsky, J., and Marcus, Y., *Ion Exchange and Solvent Extraction* (9 vols.), Marcel Dekker, New York (1966–1984).

18(c) Nachod, F., and Schubert, J., *Ion Exchange Technology*, Academic Press, New York (1956).

19(b) Naden, D., and Streat, N. (eds.), *Ion Exchange Technology*, John Wiley & Sons, New York (1984).

20(c) Osborn, G. H., *Synthetic Ion Exchangers*, Macmillan, New York (1956).

21(a) Rieman, R., and Walton, H. F., *Ion Exchange in Analytical Chemistry*, Pergammon Press, New York (1970).

22(c) Salmon, J. E., *Ion Exchange: A Laboratory Manual*, Academic Press, New York (1959).

23(b) Samuelson, O., *Ion Exchange Separations in Analytical Chemistry* (2nd ed.), John Wiley & Sons, New York (1983).

24(b) Sollner, K., Recent advances in the electrochemistry of membranes with high selectivity, *J. Electrochem. Soc.* 97, 139C (1950).

25(b) Walton, H. F. (ed.), *Ion Exchange Chromatography*, John Wiley & Sons, New York (1972).

These references can be sorted out as follows:

- 2–5 and 12–14 contain information on the practical application of ion exchange materials.
- 6, 9, 12, 21, and 23 will be helpful in the application of the ion exchange process to analytical problems.
- 17–19 and 21 have excellent general mathematical treatments for the ion phenomena associated with water treatment.
- 8, 19, and 25 are excellent references for those who are interested in the less familiar aspects of ion exchange processes and are comfortable with advanced mathematical treatments of ion exchange phenomena.
- 7, 15, 16, and 24 are related to the basic electrochemical behavior of ion exchange membranes and associated processes.

Appendix B
USEFUL ANALYTICAL METHODS

The following analytical techniques are useful, and will be referred to in the experimental units which follow later. Additional information concerning the analysis of water can be found in:

1. *Standard Methods for the Examination of Water and Waste Water*, American Public Health Association, 12th ed. (1965).
2. *Annual Book of ASTM Standards*, Vol. 11.02, "Water" (1985).

These methods can also be found in most college-level textbooks on quantitative analysis.

A TITRATION OF ACID SOLUTIONS

1. Using an appropriate pipette, take a sample of the unknown and transfer it to a clean 250 ml Erlenmeyer flask. Record the volume taken as "A" to the nearest whole milliliter.
2. Fill a 50 ml burette with the standardized NaOH solution, record the normality of the solution, and adjust the solution volume to the zero mark.
3. Add 3–4 drops of methyl orange indicator to the unknown in the 250 ml flask and mix. Slowly add the standardized NaOH solution from the burette with mixing until the solution changes color to a salmon pink color.
4. Record the titration volume as "B" to the nearest 0.1 milliliter.
5. Calculate the acid concentration as follows:

$$\text{Acidity, mg/L as } CaCO_3 = (B, \text{ ml} \times N_{NaOH} \times 50{,}000)/A, \text{ ml}$$

6. The best accuracy is obtained when the titration volume is between 20 and 40 ml. Adjust the initial sample volume "A" so that the titration volume is in the above range. As the titration endpoint is approached, the standard solution should be added dropwise so that the endpoint is not overrun.

B TITRATION OF BASIC SOLUTIONS

1. Using an appropriate pipette, take a sample of the unknown and transfer it to a clean 250 ml flask. Record the volume taken as "A" to the nearest whole milliliter.
2. Fill a 50 ml burette with the standardized H–SO solution and adjust the burette to the zero mark.
3. Add 3–4 drops of phenolphthalein indicator, and slowly add the standardized H–SO from the burette with mixing until the color of the unknown changes from a deep pink to colorless.
4. Record the titration volume as "B" to the nearest 0.1 ml.
5. Calculate the alkalinity concentration as follows:

$$\text{Alkalinity, mg/L as } CaCO_3 = (B \text{ ml} \times NH_2SO \times 50{,}000/A, \text{ ml}$$

6. The best accuracy is obtained when the titration volume is between 20 and 40 ml. Adjust the volume "A" so that the titration volume is in the above range. As the titration endpoint is approached, the standardized solution should be added dropwise to be sure the endpoint is not overrun.

C HARDNESS TITRATION WITH STANDARDIZED EDTA

1 General Discussion

1.1 Principle. When EDTA (ethylenediaminetetraacetic acid or its salts) is added to water containing both calcium and magnesium ions, it combines first with the calcium that is present. Therefore, the calcium ion can be determined directly using EDTA when the pH is made sufficiently high so that the magnesium is largely precipitated as the hydroxide and an indicator is used which combines with calcium only. Several indicators are available that will give a color change at a point where all the calcium has been complexed by the EDTA at a pH of 12–13.

1.2 Interferences. Under the conditions of this test, the following concentrations of ions cause no interference with the determination of calcium hardness:

Ferrous ion	2 mg/L
Ferric ion	20 mg/L
Manganese	10 mg/L
Zinc	5 mg/L
Lead	5 mg/L
Aluminum	5 mg/L
Tin	5 mg/L

Orthophosphate will precipitate calcium ion at the pH of this test. Strontium and barium will be titrated as calcium ion, and alkalinity in excess of 30 mg/L may cause an indistinct endpoint with hard water samples.

2   Reagents

2.1   Sodium Hydroxide 1N.   Dissolve 40 g of NaOH in a small amount of distilled water, and dilute to 1 L volumetrically with distilled water.

2.2   Indicator Solution.   Eriochrome Black T hardness indicator is available from most chemical supply houses. Preparation of a stable indicator solution in the laboratory is cumbersome and often not successful. The indicator should be purchased if possible.

2.3   Standard EDTA Solution, 0.01 M.   Weigh out 3.723 g of dry disodium salt of ethylene diamine tetra acetic acid (reagent grade), dissolve in a small amount of distilled water and dilute volumetrically to 1 L with distilled water. One milliliter of this solution is equivalent to 0.400 mg of calcium as Ca or 1.00 mg calcium as CaCO.

3   Procedure

Because of the high pH used in this procedure, the titration should be performed immediately after the addition of the sodium hydroxide.

Use a 50 ml sample volume, or smaller aliquot diluted to 50 ml, so that the calcium content is about 5–10 mg. Record the sample volume to the nearest whole milliliter. For very hard water samples with alkalinities >300 mg/L, the endpoint may be improved by taking a smaller sample and diluting to 50 ml or by neutralizing the alkalinity with acid and boiling for 1 min to remove the carbon dioxide formed.

Add 2.0 ml of the 1N sodium hydroxide solution, or a sufficient quantity to adjust the pH to 12–13. Stir and add 2–4 drops of the indicator solution. Add the standardized EDTA titrant slowly, with continuous stirring to the proper endpoint.

The calculation of total hardness is done as follows:

$$\text{T.H. ppm as } CaCO_3 = B, \text{ ml} \times 1.002 \times (1000/A, \text{ ml})$$

D   DETERMINATION OF CHLORIDE ION

1   General Description

1.1   Principle.   In a neutral or slightly alkaline solution, potassium chromate can be used to indicate the endpoint of the silver nitrate titration of chloride ion. Silver chlorine is quantitatively precipitated before any red silver chromate is formed.

1.2 Interferences. Substances in amounts normally found in potable water supplies do not interfere with this determination. Sulfite, sulfide, and thiosulfate ions when present can be eliminated as interferences by treatment with hydrogen peroxide at pH 9.0 Orthophosphate in excess of 25 mg/L interferes by precipitating as silver phosphate. Iron in excess of 10 mg/L will mask the endpoint.

2 Reagents

2.1 Chloride Free Water. For low-level analytical work, remove any chloride impurity from the distilled water by redistillation in an all-pyrex apparatus, or pass the water through a newly regenerated mixed-bed ion exchange column.

2.2 Indicator Solution. Dissolve 50 g of potassium dichromate in a little distilled water. Add a small amount of silver nitrate until a definite red precipitate is formed. Allow the solution to stand for 12 hr, filter, and dilute the filtrate to 1 L volumetrically with distilled water.

2.3 Standard Silver Nitrate Solution. Dissolve 2.396 g of $AgNO_3$ in distilled water and dilute to 1 L volumetrically. Standardize against 0.0141 N NaCl by the method described in paragraph 3.3. Store the silver nitrate solution in a brown bottle. One milliliter of this standardized silver nitrate solution is equivalent to 0.5055 mg of chloride ion 0.7052 mg chloride as $CaCO_3/ml$.

2.4 Sodium Chloride Standard. Dissolve 0.8241 g of dried (140°C) reagent grade NaCl in chloride-free distilled water and dilute to 1 L volumetrically with distilled water. Use 25 ml of this solution to standardize the silver nitrate solution prepared in paragraph 2.3.

2.5 Phenolphthalien Indicator. Dissolve 1.25 g of phenolphthalien disodium salt in distilled water and dilute to 250 ml volumetrically with distilled water.

2.6 Sodium Hydroxide 1N. Dissolve 40 g of reagent grade NaOH in some distilled water and dilute to 1 L with distilled water.

2.7 Sulfuric Acid 1N. While stirring, add 28 ml of concentrated H–SO cautiously to 700 ml of distilled water, cool, and dilute to 1 L with distilled water.

3 Procedure

Use a 100 ml sample or a suitable sample volume diluted to 100 ml. Record the sample volume as "A" to the nearest whole milliliter.

If the sample contains sulfide, sulfite, or thiosulfate, make the sample alkaline to phenolphthalien with sodium hydroxide solution. Add 1 ml of 3% hydrogen peroxide and stir for 5 min; neutralize with sulfuric acid before proceeding.

Samples in the pH range of 67–10 may be titrated directly. Adjust samples not in this range with 1N solutions of sulfuric acid or sodium hydroxide as appropriate. Add 1.0 ml of potassium dichromate indicator and mix. Titrate with the standardized silver nitrate solution to a pinkish-yellow endpoint. Record the titration volume as "B" to the nearest 0.1 ml. Carry out a blank determination starting with 100 ml of distilled water. Record the blank titration as "C" to the nearest 0.1 ml. A blank of 0.2 or 0.3 ml is usual for this method.

The concentration of Cl ion in the unknown is calculated in the following way:

$$\text{Chloride, mg/L as Cl} = (B\text{-}C) \times 0.5055 \times (1000/A, \text{ml})$$
$$\text{Chloride, mg/L as CaCO}_3 = (B\text{-}C) \times 0.7052 \times (1000/A, \text{ml})$$

E THE COLORIMETRIC DETERMINATION OF SILICA

1 General Description

1.1 Principle. Ammonium molybdate at about 1.2 pH reacts with silica and phosphates to form heteropoly acids. Oxalic acid is added to selectively remove the phosphate interference. Even when phosphates are known to be absent, the oxalic acid addition is a necessary step. The intensity of the yellow color formed is proportional to the amount of reactive silica present in the solution. In at least one of its forms, silica does not react with molybdate even though it is capable of passing through a microfine filter membrane and does not display noticeable turbidity. It is not known to what extent such "unreactive silica" occurs in water supplies, and the literature on the subject is contradictory. In the past, such terms as "colloidial," "crystalloidal," and "ionic" have been used to describe various forms of silica in natural water supplies; but such terminology has never been substantiated in practice. An optional step is included in this procedure to convert any "unreactive" silica into a "reactive" form. It must be clearly understood that these terms do not imply reactivity, or lack of it, toward other reagents or processes.

1.2 Interferences. Since apparatus and reagents may both contribute to silica, care needs to be taken to avoid the use of glassware as much as possible and to use reagents which are low in silica impurities. Also, blank determinations should be made to correct for the silica so introduced. In this method, tannin, large amounts of iron, color, turbidity, sulfide, and phosphates are potential sources of interference. The addition of oxalic acid eliminates the interference from phosphates and tannin. Photometric compensation can be used to cancel the interference due to color or turbidity when necessary.

2 Reagents

2.1 Sulfuric Acid 1N. Cautiously add 28 ml of concentrated H-SO to about 800 ml of distilled water while stirring; cool and dilute to 1 L volumetrically.

2.2 Hydrochloric Acid 1N. Add 83 ml of concentrated hydrochloric acid to 500 ml of distilled water and dilute to 1 L with distilled water.

2.3 Ammonium Molybdate Reagent. Dissolve 10 g of (NH)-Mo-O-4H-O in distilled water, warm gently, and dilute to 100 ml. Filter and neutralize to pH 7–8 with ammonium hydroxide. This solution can be stored in polyethylene indefinitely.

2.4 Oxalic Acid Solution. Dissolve 10 g of oxalic acid dihydrate in distilled water, and dilute to 100 ml.

2.5 Silica Stock Solution. Dissolve 4.730 g of reagent grade sodium metasilicate nonahydrate in recently boiled and cooled distilled water, and dilute to about 900 ml. Adjust the volume of this solution to contain exactly 1000 mg/L silica. Store this solution in a tightly sealed polyethylene bottle.

2.6 Standard Silica Solution. Dilute 10 ml of the stock solution prepared in step 2.5 to 1 L with recently boiled and cooled distilled water. Store this standard solution in a clean dry polyethylene bottle. This solution contains 0.01 mg of silica as silica per ml.

3 Procedure

To a 50 ml sample of the untreated water, add in rapid succession 1 ml of 1N HCl and 2 ml of ammonium molybdate reagent; mix and let stand for 5–10 min. Add 1 ml of oxalic acid solution and mix thoroughly; after 2 min and before 15 min read the % transmission at 640 nm. Compare the % transmission with a calibration curve prepared with the standard silica solution as described next.
    Prepare a calibration curve from a series of six standards chosen to cover the optimum range. Each standard shall be treated as described above, and the recorded % transmissions plotted against silica content on semilog paper. The concentration of silica in the unknown is determined from the calibration curve.

Appendix C
LABORATORY CHECKLIST

There are 20 experimental units in this course. This appendix provides the instructor and the student with a convenient checklist of the following:

- Equipment and instruments required for each experimental unit.
- Solutions and reagents needed for each experimental unit.

Table C-1. Checklist for Laboratory Work Units

Equipment and/or Solutions	Experimental Units
Culture slides	1,3
Glass plate	1
Test column	2,3,5,9,10,14,15,17
Addition funnel	2,3,5,14,15,17,20
Pipette(s)	2,9–12,14,18–20
Filter crucibles	4,6,13
Graduates	2,5,9,10,12
Beakers	2,4,6,7,12,18
Flasks	11–14,17
Volumetric flasks	9–11
Burette	9–12,14,17,19,20
Water aspirator	4,5
Desiccator	4,9,10,13
Drying oven	4,6,9,10,13
Magnetic Stirrer Plate	8
Balance	1,2,4,9–16,18–20
Magnifier (25×)	1,2
Water bath	13
Stopwatch	5
pH meter	7
pH electrodes	7,8
Conductivity meter	14–16
Colorimeter	14,18
Abbe refractometer	17
NaCl	11,15–17
Na_2SO_4	16
Na_3PO_4	16
NaHCO	16

Table C-1. (*Continued*)

Equipment and/or Solutions	Experimental Units
Na_2CO_3	16
CaCl	11,20
Glycerin	17
Ethylene glycol	19
Glacial acetic acid	19
Acetone	13
n-Heptane	4
Hydrogen peroxide (3%)	13
$FeSO-7H_2O$	13
Phenylene diamine–HCl	16
Isopropyl alcohol	1,10
Deionized water	1–18,20
NaCl 0.5%	12
5%	3,9 & 10
10%	3
15%	3
NaOH 2.5 meq/liter	3
4%	14
10%	10
H_2SO_4 0.8%	11
2%	14
5%	11
HCl 10%	9,10,11 & 13
NaNO 10%	7 & 10
$AgNO_3$ 1%	7,10 & 11

Standardized Solutions	
0.02 N NaEDTA	14,20
0.02 N NaOH	20
0.05 N NaOH	11,12
0.1 N NaOH	9,11,19
0.1 N NaOH[a]	9
0.1 N H_2SO_4	9–11
0.1 N HCl	18
0.1 N $AgNO_3$	10,17

Indicators	
Methyl orange	7,13
Methyl red	14
Phenolphthalein	9–12,19
Potassium chromate	10,17
Eriochrome black	14,20
Anion test water	14
Cation test water	14

[a]Prepared in 5% NaCl solution and standardized.

GLOSSARY OF ION EXCHANGE TERMS

The following terms are useful definitions associated with ion exchange technology. Many of the terms will be encountered in the experimental sections of this Manual. The student will also find many of these definitions useful when reading ion exchange publications.

Absolute Density: The weight of wet resin which occupies a unit volume, usually expressed as grams per unit volume of resin in a specified ionic form.

Activity Coefficient: A factor which, when multiplied by the active concentration, yields the total mass in solution or resin phase.

Adsorption: The removal of materials from solution by physical attraction without the exchange of ionic species.

Adsorption Isotherm: A graph showing the amount of material absorbed as a function of the equilibrium concentration at a fixed temperature per unit weight of ion exchange material.

Air Mixing: The process of mixing two ion exchange materials of different densities in a water slurry with air to yield a uniform mixed bed.

Alkalinity: The quantitative capacity of a raw water supply to react with hydrogen ions.

Anode: The electrode at which oxidation occurs; it is the positive pole for the electrodialysis process.

Anolyte: The electrolyte surrounding the anode which carries the electric current.

Attrition: Breakage and wear of ion exchange materials.

Backwashing: The upward flow of water through an ion exchange bed to clean it of foreign material, reduce the compaction of the bed, and classify the ion exchange particles.

Base Exchange: The exchange of cations between a solution and the cation exchange material. The term is most commonly used before the ionic concept was developed (see *Ion Exchange*).

Batch Contact: A method of using ion exchange materials in which the resin and liquid to be treated are mixed in a vessel and the liquid is decanted off after equilibrium is attained.

Bead Count: The evaluation of an ion exchange material's physical quality by determination of percent whole, cracked, and broken beads in a wet sample.

Bed: The amount of ion exchange material contained in a column or operating unit.

Bed Depth: The depth of ion exchange material in a column or operating unit between the inlet distributor and the effluent collector.

Bed Expansion: The fluidized rise of ion exchange particles in a column during backwashing.

Bed Volume: The volume of ion exchange material of specified ionic form contained in a column or operating unit, usually measured as the backwashed and drained volume, and expressed as cubic feet.

Bifunctional: An ion exchange material which contains two different active groups on the same polymer matrix.

Breakthrough: That volume of effluent where the concentration of the exchanging ion in the effluent reaches a predetermined limit. This point is usually the practical end of the exhaustion cycle and the beginning of the regeneration cycle.

Capacity: The number of equivalents of exchangeable ion per unit volume, unit wet weight, or dry weight of an ion exchange material in a standard ionic form.

Carbonaceous Exchangers: Cation exchange materials produced by the sulfonation of coal or other natural cabonaceous raw materials.

Cathode: The electrode at which reduction occurs; it is the negative pole for an electrodialysis process.

Catholyte: The electrolyte surrounding the cathode which carries the electric current.

Channeling: The creation of isolated paths of less resistance in an ion exchange resin bed caused by the introduction of air pockets, dirt, and other factors which produce uneven pressure gradients. Channeling prevents uniform treatment of the liquid being processed.

Chemical Stability: The ability of an ion exchange material to resist changes in its properties when in contact with aggressive chemical solutions, such as oxidizing agents. Also, the ability to resist degradation due to changes in the structure of the resin.

Chromatography: The separation of ions, molecular species, or complexes into highly purified fractions by means of ion exchange materials or adsorbents.

Clumping: The formation of agglomerations in an ion exchange bed due to fouling, or electrostatic charges.

Colorthrow: The leaching of color bodies from an ion exchange material to a process liquid.

Column Operation: The most common method of employing ion exchange materials, in which the liquid to be treated passes through a fixed bed of ion exchange resin.

Concentration History: The ratio of effluent concentration to influent concentration, C/C_0, shown as a function of the volume treated.

Concurrent Operation: Ion exchange operation in which the process water and regenerant are passed through the bed in the same direction.

Conductivity: The reciprocal of the resistance in ohms measured between opposite faces of a centimeter cube of an aqueous solution at a specific temperature.

Contact Time: The amount of time which the process liquid spends in the ion exchanger bed expressed in minutes. Determined by dividing the bed volume by the flowrate using consistent units.

Coulomb: The quantity of electricity flowing for 1 sec at a current strength of 1 ampere ($=1/96500$ Faraday $=1$ A).

Countercurrent Operation: An ion exchange column operation in which the resin and process liquid are moving at the same time in opposite directions.

Counterflow Operation: An ion exchange operation in which the process liquid and regenerant flows are in opposite directions.

Counter Ion: Anion contained in the ion exchanger matrix to preserve electroneutrality as the ion exchange process is taking place.

Cross-linking: Binding of the linear polymer chains in the matrix of an ion exchange material with an agent which produces a three-dimensional insoluble product.

Dealkalization: An anion exchange process for the removal or reduction of alkalinity in a water supply.

Deashing: The removal of inorganic ions from sugar solutions to provide a final product stream with low ash content.

Decationization: The exchange of cations for hydrogen ions by a strong acid cation exchange material in the hydrogen form (see *Salt Splitting and Cation Exchange*).

Decolorization: The removal of color from process liquids by ion exchange or absorption.

Decross-linking: The alteration of the ion exchange structure by disruption of the cross-linking by aggressive chemical attack or heat. This causes increased swelling of ion exchange materials.

Diffusion, Film: The movement of ions through the liquid film on the surface of an ion exchange particle.

Diffusion, Particle: The movement of ions within the ion exchange material toward unoccupied exchange sites.

Degradation: The physical or chemical reduction of ion exchange properties due to type of service, solution concentration used, heat, or aggressive operating conditions. Some effects are capacity loss, particle size reduction, excessive swelling, or any combination of the above.

Deionization: The removal of ionizable constituents from a solution or process liquid by ion exchange. Normally, this is achieved by passing the process liquid through the hydrogen form of the cation exchanger and then through the hydroxyl form of the anion exchanger. The reactions involved in a two-bed system are:

(a) Cation exchange

$$R_c-H + NaCl \rightarrow R_c-Na + HCl$$

(b) Anion Exchange

$$R_a-OH + NaCl \rightarrow R_a-Cl + NaOH$$

When the process is carried out in a mixed-bed system, the overall reaction is:

$$R_c-H = R_a-OH = NaCl - R_c-Na = R_a-Cl = H_2O$$

Demineralization: (See *Deionization.*)

Density: The density of an ion exchange material usually expressed as wet g/L or dry g/L.

Desalination: The removal of dissolved salts from a solution to produce either a usable water or the purification of a process liquid which has to be free of dissolved salts.

Distribution Coefficient: The ratio of the milliequivalents of ion A in the resin phase to the milliequivalents of ion A in the aqueous phase at equilibrium.

Effective Size: The particle size expressed in milliliters which represents 90% of the ion exchange material from a screen analysis.

Efficiency: Usually expressed as Kg/L or lb/ft³ as $CaCO_3$. Often expressed as the ratio of the weight of regenerant used compared to the theoretical value taken as unity.

Effluent: The process liquid which emerges from an ion exchange column or bed.

Electrodialysis: The process which separates positively and negatively charged ions by using an electric current and selective ion exchange membranes.

Eluate: The solution resulting from an elution process.

Elution: The stripping of ions or complexes from an ion exchange material by passing through the bed solutions containing other ions at specific known concentrations.

Exhaustion: The step in an ion exchange cycle in which the undesirable ions are removed from the process liquid. When the supply of desirable ions on the resin exchange sites is almost fully depleted, the ion exchange material is said to be exhausted.

Faraday: The amount of electricity required to transport one gram-equivalent of ions in an electrodialysis process.

Fast Rinse: That portion of the rinse which follows the slow rinse; usually passed through the ion exchange bed at the service flowrate.

Fines: Extremely small particles of an ion exchange material which are undesirable for a particular ion exchange operation. The presence of fines will cause high pressure drops.

Fouling: Any relatively insoluble deposit or film on an ion exchange material which interferes with the application of interest.

Fluvic Acid: A high molecular weight polycarboxylic acid often found in surface water supplies. It often contributes to organic fouling of ion exchange materials.

Freeboard: The space provided above a resin bed in a column or operating unit to accommodate the expansion of the resin bed during the backwash cycle.

Grains/Gallon: A measure of concentration equal to 17.1 mg/L.

Headloss: The loss of liquid pressure head resulting from the passage of the liquid through a bed of ion exchange material.

HETP: Height Equivalent of a Theoretical Plate. Determined by evaluating the various processes taking place in an ion exchange column.

$$\text{HETP} = H_s + H_p + H_f + H_l$$

where H_s is the contribution due to particle size; H_p is the contribution due to particle diffusion; H_f is the contribution due to film diffusion; H_l is the contribution due to lateral liquid phase diffusion.

Humic Acid: A high molecular weight polycarboxylic acid found in surface water supplies which contributes to organic fouling in ion exchange materials.

Hydrogen Cycle: A cation operation in which the regenerated form of the ion exchange material is the hydrogen form.

Hydroxide Cycle: An anion exchange operation in which the regenerated form of the ion exchange material is the hydroxyl form.

Ionization: The separation of part or all of the solute molecules into positive (cationic) and negative (anionic) ions in a dissociating media such as water.

Ion Exclusion: A process in which ionized species are separated from nonionized or weakly ionized species using an ion exchange material.

Ion Retardation: The technique of separating strong electrolytes from weak electrolytes in which the ion exchange material acts as an absorption media.

Influent: The solution or process liquid entering an ion exchange column.

Interstitial Volume: The space between the particles of an ion exchange material in a column or an operating unit (see *Void Volume*).

Layered Bed: Commercial application in which two ion exchange materials (i.e., weak and strong base anion resin) are contained as undisturbed layers in a single operating unit.

Leakage: The appearance of ions in the effluent stream which should be removed by the ion exchange process. Leakage may be due to unfavorable equilibria, incomplete regeneration, high process flowrates, channeling, or a combination of the above factors.

Membrane: A thin sheet separating different streams, which contains active groups that have a selectivity for anions or cations but not both.

Ohm: The unit of resistance of a solution, often related to the electrolyte concentration.

Operating Cycle: A complete ion exchange process consisting of a backwash, regeneration, rinse, and service run.

Osmotic Stability: The ability of an ion exchange material to resist physical degradation due to volume changes imposed by repeated applications of dilute and concentrated solutions.

Permeability: The ability of an ion exchange membrane to pass ions under the influence of an electric current.

Permselectivity: The ability of an ion exchange membrane to the selective passage of anions or cations under the influence of an electric current.

pH: The negative logarithm (base 10) of the hydrogen ion concentration in water.

pK: The negative logarithm (base 10) of the equilibrium constant K for an ion exchanger in a dissociating media such as water.

Physical Stability: The ability of an ion exchange material to resist breakage caused by mechanical manipulation.

Polisher: A mixed-bed ion exchange unit usually installed after a two-bed deionizer system to remove the last traces of undesirable ions.

Pressure Drop: (See *Headloss.*)

ppb: Unit of concentration, parts per billion equal to 1 $\mu g/L$.

ppm: Unit of concentration, parts per million equal to 1 mg/L.

Redox Resin: An ion exchange material which is capable of taking part in reversible oxidation-reduction reactions by electron transfer.

Regenerant: The solution used to convert an ion exchange material from its exhausted state to the desired regenerated form for reuse.

Regeneration: The displacement from the ion exchange material of the ions removed during the service run. Performed by passing the regenerant through the bed.

Regeneration Level: The amount of regenerant chemical used per unit volume of ion exchanger bed, commonly expressed as lb/ft^3 or Kg/m^3.

Rinse: The passage of water through an ion exchange material to remove excess regenerant. (See *Fast rinse* and *Slow rinse.*)

Reverse Demineralization: A demineralization process in which the process liquid passes first through the regenerated anion exchange material before passing through the regenerated cation exchange material.

Salt-Splitting: The conversion of neutral salts to their corresponding acids or bases by passing their solutions through strong acid or strong base ion exchange materials.

(a) Cation Exchange Salt-Splitting

$$R_c-H + NaCl \rightarrow R_c-Na + HCl$$

(b) Anion Exchange Salt-Splitting

$$R_a-OH + NaCl \rightarrow R_a-Cl + NaOH$$

Scavenger: A polymer matrix or ion exchange material used to specifically remove organic species from the process liquid before the solution is deionized.

Selectivity: Is a quotient equal to the ratio of ions A and B in the resin phase to the ratio of ions A and B in the aqueous phase at equilibrium in accordance with the reaction:

$$K_B^A = \frac{(A)_r}{(B)_r} \times \frac{(B)_s}{(A)_s}$$

Service Run: (See *Operating Cycle*.)

Slow Rinse: That portion of the rinse which follows the regenerant solution and is passed through the ion exchange material at the same flowrate as the regenerant.

Strong Acid Capacity: That part of the total cation exchange capacity which is capable of converting neutral salts to their corresponding acids. (See *Salt-Splitting*.)

Strong Acid Cation Exchanger: A cation exchange material with an active group capable of splitting neutral salts to form their corresponding free acids. (See *Salt-Splitting*.)

Strong Base Capacity: That part of the total anion exchange capacity capable of converting neutral salts to their corresponding bases. (See *Salt-Splitting*.)

Strong Base Anion Exchanger: An anion exchange material with an active group capable of splitting neutral salts to form their corresponding free bases. (See *Salt Splitting*.)

Sweetening Off: The portion of an ion exchange operating cycle that is necessary to bring the effluent concentration down to a level equal to the water used for the backwash process.

Sweetening On: The portion of an ion exchange operating cycle required to bring the effluent concentration of the process liquid up to that of the influent.

Total Capacity: The ultimate ion exchange capacity of any ion exchange material.

Transport Number: In electrodialysis, a number indicating the fraction of the total electric current carried by a specific ionic species. If n is the anionic transport number, then $1 - n$ is the cationic transport number.

Uniformity Coefficient: The volume or weight ratio of the 90% retention size (see *Effective Size*) and the 40% retention size in a screen analysis.

$$\text{Uniformity coefficient} = \frac{\text{Opening retaining 90\%, mm}}{\text{Opening retaining 40\%, mm}}$$

Void Volume: (See *Interstitial Volume.*)

Water Regain: (See *Water Retention.*)

Water Retention: The amount of water, expressed as a percent of the wet weight, retained within and on the surface of a fully swollen and drained ion exchange material.

Weak Acid Cation Exchangers: Those cation exchange materials with groups which cannot split neutral salts to form the corresponding free acids.

Weak Base Anion Exchangers: Those anion exchange materials with groups which cannot split neutral salts to form the corresponding free bases.

Zwitterion: An ion carrying both positive and negative charges—thus, being electrically neutral. Molecules of this type may be highly polar depending on the charge distribution.

Appendix E

TABLES AND CONVERSION FACTORS

The tables given in this section have been taken from several sources. All will be useful to individuals concerned with the field of ion exchange technology. They are included here as an aid to the student and instructor during the use of the laboratory manual. [*Sources:* Tables E-1 and E-6 to E-8, Lange's *Handbook of Chemistry*; Tables E-2 to E-5, *Duolite Ion Exchange Manual*; Table E-9, courtesy of SYBRON Chemical Co.]

Table E-1. Conversion Units Useful in Ion Exchange Calculations

To Convert	To	Multiply By
Atmosphere	Inches of Hg	29.921
	Pounds per in.^{2}	14.696
Angstrom (A)	Inches	3.973×10^{-9}
	Meters	1.000×10^{-10}
	Microns	1.000×10^{-4}
Bar	Atmosphere	0.99869
Baume	Specific gravity	
Centimeter (cm)	Inches	0.3937
	Angstroms	1.000×10^{8}
Centimeters/sec	Feet/sec	3.281×10^{-2}
Cubic centimeters	Cubic feet	3.531×10^{-5}
	Gallons (U.S.)	2.642×10^{-4}
Cubic foot	Cubic meters	2.832×10^{-2}
	Gallons (U.S.)	7.4805
	Liters	28.32
Cubic feet/min	Liters/sec	0.472
Cubic inches	Cubic centimeter	16.43
	Gallons (U.S.)	4.329×10^{-3}
	Liters	1.639×10^{-2}
Cubic meter	Cubic feet	35.31
	Gallons (U.S.)	264.2
Feet	Meters	0.3048
Feet/min	Centimeters/sec	0.508

Table E-1. (*Continued*)

To Convert	To	Multiply By
Gallons (U.S.)	Pounds, H O	8.3378
	Cubic centimeters	3.785×10^3
	Cubic feet	0.1337
	Liters	3.785
	Gallons (British)	0.83268
Gallons/min	Liters/sec	6.308×10^{-2}
	Cubic feet/sec	2.228×10^{-3}
Gallons/ft^3	Liters/m^2	1.336×10^2
Gallons/ft^2	Liters/in.^{2}	2.628×10^{-2}
Grains/gallon (U.S.)	mg/liter, ppm	17.118
	Pounds	2.2046×10^{-3}
	Grains/gal (Troy)	15.431
Grams/cm^3	Pounds/gal (U.S.)	8.345
Grams/L	Pounds/1000 gal. (U.S.)	8.345
Inches	Centimeters	2.540
	Angstroms A	2.540×10^8
kg/ft^3 as CaCO	Meq/L	45.8
Kilogram	Pounds	2.2046
Kilogram/m^3	Pounds/ft^3	6.243×10^{-2}
Kilogram/m^2	Pounds/ft^2	0.2048
Liter	Cubic feet	3.531×10^{-2}
Liters/min	Gallons/sec	4.404×10^{-3}
Meters	Angstroms (A)	1.000×10^{10}
	Feet	3.281
	Inches	39.37
Microns	Centimeters	1.000×10^4
	Inches	3.987×10^{-5}
Mgs/L as CaCO	Equivalents/L	2.000×10^{-5}
MHO	Ohm	1/MHO
Pounds	Grams	4.5359×10^2
Pounds (H O)	Gallons (U.S.)	0.1198
Pounds/ft^3	Kg/m^3	16.02
Pounds NaCl/ft^3	Gram equiv./L	0.274
Pounds HCl/ft^3	Gram equiv./L	0.439
Pounds H SO/ft^3	Gram equiv./L	0.327
Pounds NaOH/ft^3	Gram equiv./L	0.400
Square centimeters	Square meters	1.000×10^{-4}
	Square feet	1.076×10^{-3}
Square feet	Square meters	9.29×10^{-2}
Square inches	Square feet	6.944×10^{-3}
Square meters	Square feet	10.76
Tons (metric)	Kilograms	1.000×10^3
Tons (short)	Pounds	2.000×10^3
Tons (short)	Tons (metric)	0.9078

Table E-2. Conversion of Water Hardness Units

Old Expression	Equivalent (ppm as CaCO)
1 grain CaCO/gallon (U.S.)	17.118
1 grain CaCO/gallon (British)	14.29
1 part CaCO/$1 \times 10^{+5}$ parts water (French)	10.00
1 part CaO/$1 \times 10^{+5}$ parts water (German)	17.85
1 part CaO/$7.14 \times 10^{+5}$ parts water (Russian)	2.50

Table E-3. Conversion of Weights

$$
\begin{aligned}
1 \text{ gram} &= 1000 \text{ milligrams} \\
1 \text{ kilogram} &= 1000 \text{ grams} \\
&= 2.2046 \text{ pounds} \\
1 \text{ milligram} &= 1000 \text{ micrograms} \\
1 \text{ microgram} &= 1000 \text{ nanograms} \\
1 \text{ metric ton} &= 1000 \text{ kilograms} \\
&\quad\ 2204.6 \text{ pounds} \\
1 \text{ gram} &= 15.4324 \text{ grains} \\
&= 2.204 \times 10^{-3} \text{ pounds} \\
1 \text{ grain} &= 6.480 \times 10^{-2} \text{ grams} \\
&= 1.429 \times 10^{-4} \text{ pounds}
\end{aligned}
$$

Table E-4. Conversion of Concentrations

$$
\begin{aligned}
1 \text{ ppm} &= 1 \text{ mg/L} \\
&= 1/120 \text{ lb}/1000 \text{ gal (U.S.)} \\
1 \text{ g/L} &= 1000 \text{ mg/L} \\
&= 6.243 \times 10^{-2} \text{ lb/ft}^3 \\
1 \text{ mg/L} &= 1000 \ \mu g/L \\
1 \text{ lb/ft}^3 &= 16.0189 \text{ g/L} \\
1 \text{ g/cm}^3 &= 62.426 \text{ lb/ft}^3 \\
&= 8.334 \text{ lb/gal (U.S.)}
\end{aligned}
$$

Table E-5. Conversion of Volumes

$$
\begin{aligned}
1 \text{ milliliter} &= 1.000 \times 10^{-3} \text{ L} \\
&= 1.000 \times 10^{-6} \text{ m}^3 \\
&= 2.642 \times 10^{-4} \text{ gal (U.S.)} \\
1 \text{ liter} &= 1000 \text{ ml} \\
&= 0.26418 \text{ gal (U.S.)} \\
1 \text{ gallon (U.S.)} &= 3785 \text{ ml} \\
&= 3.785 \text{ L} \\
&= 0.83268 \text{ gal (British)} \\
1 \text{ cubic foot} &= 28316 \text{ ml} \\
&= 28.316 \text{ L} \\
&= 28316 \text{ cm}^3 \\
&= 7.48\text{m gal (U.S.)}
\end{aligned}
$$

Table E-6. Conversion of Flowrates

$$
\begin{aligned}
1 \text{ L/min} &= 0.2642 \text{ gal (U.S.)/min} \\
&= 16.67 \text{ ml/sec} \\
1 \text{ m}^3/\text{min} &= 4.4029 \text{ gal (U.S.)/min} \\
&= 16.665 \text{ L/min} \\
&= 2.778 \times 10^{+2} \text{ ml/min} \\
1 \text{ U.S. gal/min} &= 0.227 \text{ m}^3/\text{hrl} \\
&= 3.785 \text{ L/min} \\
&= 6.310 \times 10 - 2 \text{ L/sec} \\
&= 63.1 \text{ ml/sec} \\
1 \text{ ft}^3/\text{min} &= 28.316 \text{ L/min} \\
&= .719 \times 10 + 2 \text{ ml/sec} \\
&\quad 4.719 \times 10^2 \text{ ml/min} \\
1 \text{ U.S. gal/ft}^2 &= 40.727 \text{ L/m}^2 \\
1 \text{ U.S. gal/ft}^3 &= 133.75 \text{ L/m}^3
\end{aligned}
$$

Table E-7. Openings for U.S. Standard Sieves

Mesh Size or Sieve Number	Opening Millimeters	Inches
10	2.00	0.0787
12	1.68	0.0661
14	1.41	0.0555
16	1.19	0.0469
18	1.00	0.0394
20	0.84	0.0331
25	0.71	0.0280
30	0.59	0.0232
35	0.50	0.0197
40	0.42	0.0165
45	0.35	0.0138
60	0.250	0.0098
70	0.210	0.0083
80	0.177	0.0070
100	0.149	0.0059
200	0.074	0.0029
325	0.044	0.0017

Table E-8. Chemical Equivalent Weights and Conversion Factors

Substance	Equivalent Weight	Substance to $CaCO_3$ Equivalent	$CaCO_3$ to Equivalent Substance
Compounds			
Aluminum chloride	44.4	1.13	0.89
Aluminum hydroxide	26.0	1.92	0.52
Aluminum sulfate	57.0	0.88	1.14
Ammonium chloride	53.5	0.94	1.07
Ammonium hydroxide	35.1	1.43	0.70
Ammonium sulfate	66.1	0.76	1.32
Barium carbonate	98.7	0.51	1.97
Barium chloride	104.4	0.48	2.09
Barium hydroxide	85.7	0.59	1.71
Barium sulfate	116.7	0.43	2.33
Calcium bicarbonate	81.1	0.62	1.21
Calcium carbonate	50.0	1.00	1.00
Calcium chloride	55.5	0.90	1.11
Calcium hydroxide	37.1	1.35	0.74
Calcium hypochlorite	71.5	0.70	1.48
Calcium sulfate	68.1	0.74	1.36
Calcium nitrate	82.1	0.61	1.64
Calcium phosphate	51.7	0.97	1.03
Copper sulfate	80.0	0.63	1.59
Ferrous carbonate	57.9	0.86	1.16
Ferrous hydroxide	44.9	1.11	0.90
Ferrous sulfate	76.0	0.66	1.52
Ferric chloride	54.1	0.93	1.08
Ferric sulfate	66.7	0.75	1.33
Magnesium bicarbonate	78.2	0.68	1.46
Magnesium carbonate	42.2	1.19	0.84
Magnesium chloride	47.6	1.05	0.95
Magnesium hydroxide	29.2	1.71	0.58
Magnesium nitrate	74.2	0.67	1.48
Magnesium phosphate	43.8	1.14	0.88
Magnesium sulfate	60.2	0.83	1.20
Manganese chloride	62.9	0.80	1.25
Manganese hydroxide	44.4	1.13	0.89
Potassium carbonate	69.1	0.72	1.38
Potassium chloride	74.6	0.67	1.49
Potassium hydroxide	56.1	0.88	1.12
Sodium bicarbonate	84.0	0.60	1.68
Sodium carbonate	53.0	0.94	1.06
Sodium chloride	58.5	0.85	1.17
Sodium hypochlorite	74.5	0.67	1.49
Sodium hydroxide	40.0	1.25	0.80
Sodium nitrate	85.0	0.59	1.70
Sodium nitrite	69.0	0.73	1.38
Trisodium phosphate	54.7	0.91	1.09
Disodium phosphate	71.0	0.70	1.42
Monosodium phosphate	120.0	0.42	2.40
Sodium sulfate	71.0	0.70	1.42

Table E-8. Chemical Equivalent Weights and Conversion Factors

Substance	Equivalent Weight	Substance to $CaCO_3$ Equivalent	$CaCO_3$ to Equivalent Substance
Acids			
Acetic acid	60.1	0.83	1.20
Carbonic acid	31.0	1.61	0.62
Hydrochloric acid	36.5	1.37	0.73
Nitric acid	63.0	0.79	1.26
Phosphoric acid	32.7	1.53	0.65
Sulfuric acid	49.0	1.02	0.98
Sulfurous acid	41.1	1.22	0.82
Anions			
Bicarbonate	61.0	0.82	1.22
Carbonate	30.0	1.67	0.60
Chloride	35.5	1.41	0.71
Fluoride	19.1	2.61	0.38
Nitrate	62.0	0.81	1.24
Nitrite	46.0	1.02	0.92
Phosphate	31.7	1.58	0.63
Phosphate (dihydrogen)	97.0	0.52	1.94
Phosphate (mono-hydrogen)	48.0	1.04	0.96
Silicate	38.1	1.31	0.76
Sulfate	48.0	1.04	0.96
Sulfite	40.0	1.25	0.80
Cations			
Aluminum	9.0	5.56	0.18
Ammonium	18.9	2.78	0.36
Barium	68.7	0.73	1.37
Calcium	20.0	2.50	0.40
Copper	31.8	1.57	0.64
Divalent iron	27.9	1.79	0.56
Trivalent iron	18.6	2.69	0.37
Hydrogen	1.0	50.0	0.02
Magnesium	12.2	4.12	0.25
Divalent manganese	27.5	1.82	0.55
Trivalent manganese	18.3	2.73	0.37
Potassium	39.3	1.28	0.78
Sodium	23.0	2.18	0.46
Zinc	32.7	1.54	0.65

Table E-9. Approximate Electrolyte Concentration Versus Specific Conductivity and Resistivity for Water at 25°C

Conductivity microohms	Resistivity ohms	Approximate Concentration mg/L as					
		NaCl	NaOH	HCl	CO	H SO	NaSO
0.055	18,200,000	Theoretically pure water					
0.063	16,000,000	0.025[a]	—	—	0.032[a]	0.043[a]	0.28[a]
0.071	14,000,000	0.029[a]	—	—	0.037[a]	0.044[a]	0.29[a]
0.083	12,000,000	0.034[a]	—	—	0.044[a]	0.046[a]	0.30[a]
0.1	10,000,000	0.04	—	0.01	0.05	0.05	0.32
0.2	5,000,000	0.08	0.03	0.02	0.1	0.06	0.40
1.0	1,000,000	0.4	0.2	0.13	0.8	0.16	1.00
2.0	500,000	0.8	0.4	0.26	2.5	0.27	1.74
4.0	250,000	1.6	0.8	0.55	9.5	0.51	3.21
6.0	166,000	2.5	1.0	0.9	20	0.75	4.7
8.0	125,000	3.2	1.5	1.2	40	0.98	6.2
10	100,000	4.0	2.0	1.5	70	1.2	7.7
20	50,000	8.0	4.0	2.0	320	2.4	15
30	33,300	14	5.0	3.0	730	3.6	23
40	25,000	19	6.0	4.0	1400	4.8	30
50	20,000	24	7.0	4.5	2200	6.0	38
60	16,670	28	9.0	5.5	—	7.1	45
70	14,290	33	10.5	6.5	—	8.3	53
80	12,500	38	11	7.5	—	9.5	60
90	11,110	43	13	8.0	—	11	67
100	10,000	50	14	9.0	—	12	75
200	5,000	100	27	18	—	24	149

[a]These calculated values have never been verified by direct chemical analysis.

SAMPLING ION EXCHANGE RESINS

Since ion exchange products are composed of discrete particles, care must be taken to assure that the sample represents the production lot, or ion-exchange unit in the field. Additional information concerning sampling can be found in *Annual Book of ASTM Standards*, Vol. 11.02, "Water" (1985).

The procedure described here is used most frequently to sample "as-received" products in their original shipping containers or drums, as a part of good quality-control practices.

Once the sample is obtained, it should be transferred to a gas proof, sealable container as soon as possible. A final sample volume of 1 L (1 quart) is usually sufficient for all analytical work and will provide a "file sample" for reference or rechecks.

SAMPLING "AS-RECEIVED" SHIPMENTS OR PRODUCTION LOTS

Dry or Free-Flowing Products

Before opening, the drum or container should be rocked slightly from side-to-side to assure uniform packing.

Insert the closed sampler (shown in Figure F-1) to the bottom of the container. Open and close the sampler several times. Withdraw the closed sampler and transfer the sample to a suitable, labeled container. Take at least three samples uniformly spaced on a circle two thirds the diameter of the drum or container.

For a production lot, not less than 10% of the drums or containers will be sampled as described above.

Sample volumes can be reduced to an appropriate volume by quartering as described on page 224.

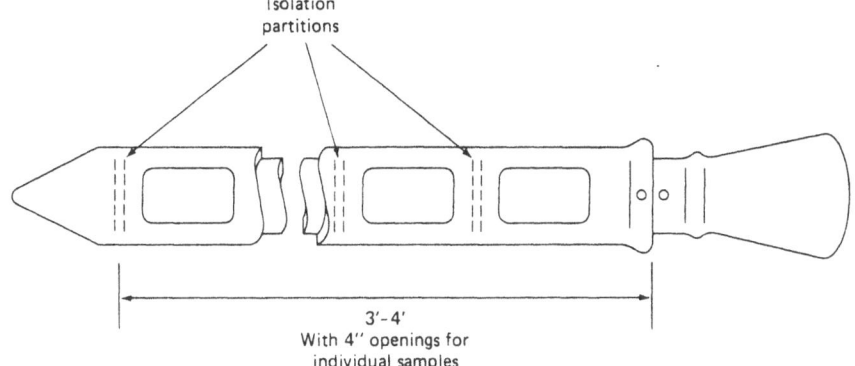

Figure F-1. Sampling device (grain thief) for dry products and mixed-bed products.

Moist Ion Exchange Products

When the moist "as-received" product is contained in a drum or fiber pack, upend the container and let it stand overnight to redistribute the excess water before sampling.

After standing overnight, right the container and rock slightly from side-to-side to assure uniform packing.

Using the sample tube (shown in Figure F-2), take at least three core samples by inserting the tube gently to the bottom of the container and withdrawing slowly. These samples should be uniformly spaced on a circle about two thirds the diameter of the container.

Transfer each core sample immediately to a suitable container and seal promptly to preserve the moisture content.

For a production lot, not less than 10% of the drums or containers will be sampled as described above.

Sample volumes can be reduced to an appropriate volume by quartering as described on page 224.

SAMPLING OPERATING UNITS IN THE FIELD

Fixed-Bed Units with One Resin Type

This technique is useful for obtaining representative samples from ion exchange units in the field. Samples can be taken at any time in the operating cycle to evaluate performance, regeneration, rinse, backwash efficiency, etc.

Drain the excess water from the ion exchange unit until the level is at the surface of the bed. At least three uniformly spaced samples will be taken through the top manhole.

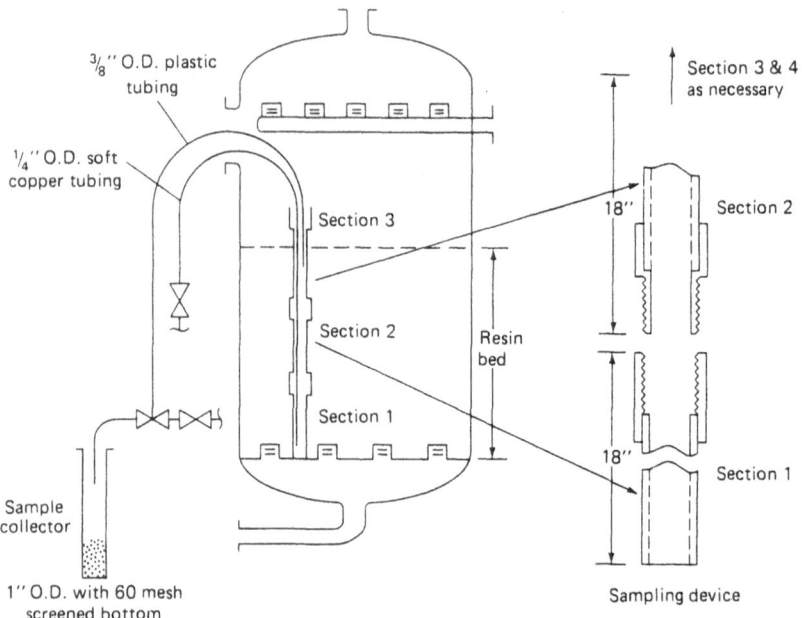

Figure F-2. Field sampling with unrestricted headroom.

Gently thrust the sampling rod (show in Figure F-2) down through the resin bed until it is contact with the bottom of the vessel or the support media.

Using a gentle up-and-down motion, slip the sample tube (shown in Figure F-2) over the rod and down through the resin bed until it is seated on the tapered stopper.

Insert the handle in the rod and pull to withdraw the sample from the ion exchange unit. The wet sample is transferred to a suitable container. After settling, the excess water can be decanted and discarded.

Sample volumes can be reduced to an appropriate volume by quartering as described on page 224.

Fixed-Bed Units with Mixed-Bed Resins

Mixed-bed ion exchange units are often sampled to determine the anion/cation resin ratios as a function of bed depth after backwashing or mixing.

After opening the unit, drain the excess water, leaving about 3 in. of water above the bed surface. Insert the closed sampling device (shown in Figure F-1). Open and close the sampling device several times. Close and withdraw the sampler.

A labeled sample container will be required for each sample section; care should be taken to keep each sample separate and appropriately labeled. The depth of each sample is an important parameter of this technique.

SAMPLING A FIXED-BED WITH RESTRICTED HEADROOM

The sampling device used consists of a series of short sections of thin-wall tubing with threaded ends (as shown in Figure E-3). These sections are inserted in the resin bed one at a time and threaded together until the bottom of the unit or the support media is contacted.

The eductor, tubing, deionized water supply, and sample receiver is assembled (as shown in Figure F-3). The $\frac{1}{4}$ in. deionized water line is inserted to the bottom of the sampling tube. The $\frac{3}{8}$ in. sample line is inserted to a point about 6 in. below the resin bed surface.

The eductor discharges to the sample receiver, which is equipped with a bottom screen (60 mesh) to allow excess water to drain off in an unrestricted manner.

Start the eductor and the deionized water flow. Slowly insert the $\frac{3}{8}$ in. sampling line and continue until all the resin contained in the sampling tube has been transferred to the resin receiver.

Shut off all flow, remove all lines, and relocate the sample tube for a second core sample if necessary. The drained ion exchanger sample is transferred to a suitable container and labeled.

Sample volumes can be reduced to an appropriate volume by quartering as described on page 224.

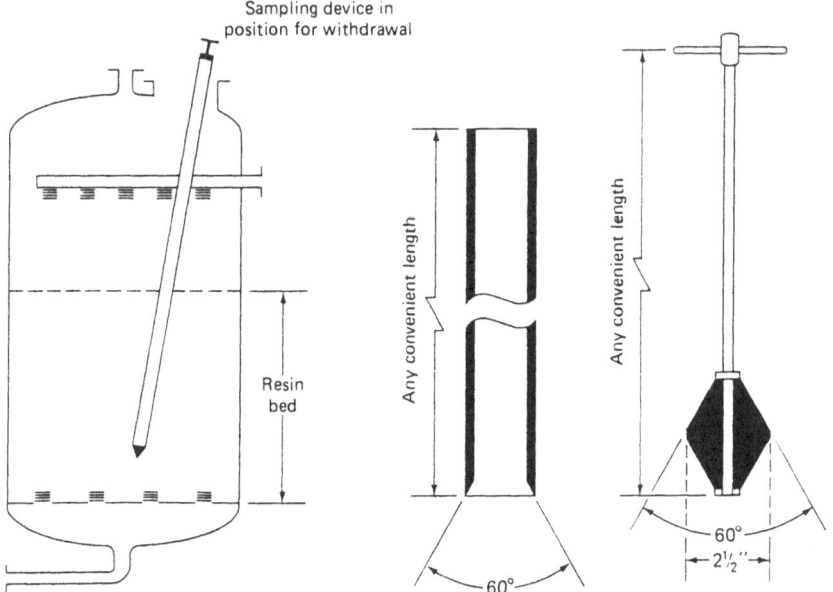

Figure F-3. Field sampling with restricted headroom.

SAMPLE SIZE REDUCTION BY QUARTERING

The following procedure is used to reduce large volume samples of particulate materials to smaller volumes while retaining the "representative" quality of the original sample.

Transfer the sample to a square polyethylene sheet, and flatten the sample into a circular shape about $\frac{3}{4}$ or 1 in. deep.

Grasp the corners of the sheet firmly, and lift to reform the sample pile.

Repeat the above steps several times to mix the sample uniformly.

Flatten the sample into a circular shape. Using a straightedge, separate the sample into four quarters, and discard the opposite quarters.

Repeat the above steps until the sample has been reduced to an appropriate volume for the required analytical work.

INDEX